W9-BWE-027
3 9153 00941874 1

TECHNOPOLIS

Social Control of the Uses of Science

NIGEL CALDER

SIMON AND SCHUSTER · NEW YORK

Published in the United States by Simon and Schuster
Rockefeller Center, 630 Fifth Avenue
New York, New York 10020

First U.S. printing

SBN 671-20496-3
Library of Congress Catalog Card Number: 79-101867
Manufactured in the United States of America

Contents

LIST OF TABLES

Let me arrest thy thoughts; wonder with mee,
Why plowing, building, ruling and the rest,
Or most of those arts, whence our lives are blest,
By cursed Cains *race invented be,*
And blest Seth *vext us with Astronomie.*
Ther's nothing simply good, nor ill alone,
Of every quality Comparison
The onely measure is, and judge, Opinion.

JOHN DONNE (1601): The Progress of the Soule

Part I

SO TECHNICALLY SWEET

ROBERT OPPENHEIMER, the physicist, was persecuted by his countrymen for alleged lack of zeal in developing the H-bomb. His martyrdom was traumatic for the American scientific community as well as for himself and it cast doubt on the possibility of ever arresting the headlong development of new weapons. Equally significant, though less publicized, was Oppenheimer's change of mind, even before his 'trial', in favour of the super-bomb when a much improved design emerged. 'Sweet and lovely and beautiful,' he judged the new idea (Davis, 1968).* It was so technically sweet it had to be tried.

How can poetic urges to change the world, using science, be made subject to the desires of ordinary people?

This first part introduces the theme. Chapter 1 attributes weakness in political control to no scientific conspiracy, but to the widespread belief that the uses of science are politically boring. In the second chapter we see patrons, from Peter the Great to Charles de Gaulle, harnessing science to nationalism

* This is the only footnote. Dates in parentheses, for example (Davis, 1968) or Lundberg (1961), refer to the bibliography.

—thus creating competing arsenals and supermarkets.

Then some dangerous and paradoxical errors must be exploded: in Chapter 3, the notion that scientists are reckless about the fate of the world; in Chapter 4, the hope that the course of research can be controlled; in Chapter 5, the fear that the uses of science can *not* be controlled. The ethical system of 'pure' science is contrasted with the lack of an equivalent system for applying science.

CHAPTER 1

Loss of Control

A NEW shape appeared in the sky over France in March 1969, the bird-like Concorde. The first prototype of the second supersonic airliner (the Russians were a few weeks ahead) climbed from the runway at Toulouse and then, in a planned anti-climax, circled at subsonic speed. André Turcat, the test pilot, had to check that the chimera of British and French design and construction would stay airborne. The people of Istre did not need that day to brace themselves for the tests of sonic boom. The skill of two nations had produced a machine of grace and power, that would cross the Atlantic Ocean like a bullet. In the Concorde, businessmen would crash the sound barrier, the idea of which, twenty years earlier, had dismayed the aces.

Yet hardly anybody wanted the Concorde. By the time of its first cautious flight, the two governments had committed more than one billion dollars to the project; once again, the estimates of an aviation project had proved ludicrously optimistic. In 1964, the incoming Labour government in Britain had wanted to pull out of the project. The airline managers who put their names down for Concordes winced at the problems of operation, economy and safety they foresaw, but they knew that some of their competitors inevitably would adopt the new machine. Controversy about the sonic boom—assuming that civilized nations would permit supersonic overflights—was compounded by the growing realization that efforts to control take-off noise in the acoustic wilderness around airports could be defeated by the newcomer's Olympus engines.

The engineers wanted the Concorde, of course, and so did the airframe and engine manufacturers. Believers in the nations' needs for technological progress of whatever kind could swallow their reservations; so could those who thought the speed of light, not the speed of sound, the only natural limit to flight. Men who thought they really needed to cross the Atlantic in three hours, might look forward to being passengers. In time, perhaps, everyone would want it. Advertising copywriters would make the Concorde the symbol of a still more modern age, passengers would flock to it and, with luck, the manufacturers might win those two hundred orders that they needed. Even old people and babies might grow accustomed to the sonic boom. Those who opposed the project would come to seem like the old fogies who contested the railway and the automobile—or would they? Even if the Concorde story might have a happy ending, no one had reason to be happy about beginnings. Whatever the outcome, the enterprise flowed from a breakdown of control of innovation.

Before the race for the supersonic airliner, weapons development had shown its own competitive logic, behind a wall of secrecy that also blocked political control. The lists of abandoned aircraft and missile projects seemed to testify to a technology restrained only by its own shortcomings. The most bewildering aspect of the post-war arms race was the accumulation of nuclear bombs by the USA during the 1950s. The US Atomic Energy Commission was required to provide the Department of Defense with all the nuclear weapons it might order, but to supply them without charge. The Department of Defense thus had an item in its arsenal for which it had to pay nothing out of its own budget; it was therefore happy to order just as many bombs as could be made. The Atomic Energy Commission, for its part, was happy to enlarge its scale and influence by developing its bomb-making resources, secure in the knowledge that the funds would be provided to meet Defense Department needs. It was a formula for a runaway process, and eventually it required presidential intervention to stop it (Encke, 1967). Thus 'overkill' came about, not by expert willfulness or wickedness, but by an elementary error of government.

Nor, as far as the engineers were concerned, was the decision to develop the Concorde taken lightly. In Britain, many years of aerodynamic research at the Royal Aircraft Establishment and the National Physical Laboratory gave good general knowledge of suitable wing forms at high speeds. Studies by the Natural Gas Turbine Research Establishment, by the aircraft manufacturers and operators, and by those responsible for air traffic control, preceded the Ministry of Aviation's establishment of a supersonic transport aircraft committee, at the end of 1956. After thirty months the committee concluded that there was a good chance that the operating costs of a supersonic machine would not be very different from those of subsonic aircraft. More than 10,000 aerodynamic models were tested in wind tunnels, and by 1960 the choice had fallen on a slender delta aircraft to be built of aluminium alloy and to fly at 2·2 times the speed of sound.

In the following year there were intensive discussions between British and French officials about the possibility of a joint development by the two countries. The French had evolved their own ideas but, by March 1962, these had been reconciled with those of the British designers. Then the conditions of Anglo-French collaboration were announced; the agreement was binding on both sides, as Harold Wilson was to rediscover six years later.

In contrast with the careful engineering work that preceded the Concorde project, parliamentary and public discussion of its desirability was virtually nil. The director of Sweden's Aeronautical Research Institute, Bo Lundberg (1961), had itemized likely snags in an article published in London while the Anglo-French talks were still in progress, but nobody paid much attention. At first sight, it looked as if the British and French aviation experts, and manufacturers in search of contracts, pressed for their project regardless of the public interest and, in conspiracy with sympathetic ministers and civil servants, sidestepped the processes of good government.

This picture is not entirely false, but it omits a revealing detail. The aerodynamicists at the Royal Aircraft Establishment who developed the concept of the Concorde foresaw the problem of the sonic boom perfectly well. In the very early days they wanted to organize a systematic series of supersonic over-flights by military aircraft to test public reaction. The Ministry of Aviation rejected the idea on the grounds that an adverse result would not be acceptable. Not until 1967 were the first rudimentary tests made in Britain.

In other words, the decision to build the Concorde was taken on political grounds, with studied disregard not only for social considerations but also for technical advice about those social considerations. Do not blame the engineers for their enthusiasm for the project, and their conscientious execution of it; in the age of the specialist, that was their job. Any fault must be laid squarely with the ministers concerned, in London and Paris, and indirectly with their cabinet colleagues who acquiesced and their parliamentary colleagues who were unwatchful.

That the time was ripe, technically, for the supersonic airliner was shown by the almost simultaneous development of the Soviet Tupolev 144, very like the Concorde. The Americans waited for something bigger and faster. At the end of 1966, the Boeing swing-wing design was chosen, but it was later judged unworkable. Meanwhile, public argument intensified about the sonic boom and other problems—such as the fear that water vapour, released by the machines into the normally dry stratosphere, might alter the climate. It eroded the simple-minded official view that the United States must meet the Anglo-French and Soviet 'challenge'. Robert Kennedy, before his assassination, wanted the supersonic airliner postponed in favour of social welfare programmes. Richard Nixon's secretary of the treasury was quick to make known his opposition to the project. Technological hesitations, after the engineers had over-reached themselves, thus gave Americans time to think what they were doing.

LEGISLATORS' FOLLIES

As the watchdogs of public affairs, parliamentarians are particularly open to charges of negligence if technology seems to be following a mindless course. Few members of the world's parliaments are very interested, let alone well informed, in scientific and technical matters; a member who is actually a qualified scientist is a curiosity. The typical parliamentarian is elderly, arts-trained, unaccustomed to sustaining deep thought, and equipped with political rules of thumb scarcely more relevant to shaping affairs today than the theory of phlogiston is to modern chemistry. After the 1965 general election, when the British press was greeting a new intake of bright young members of parliament, a check of the biographies showed only one scientific research worker—a physicist from Manchester. Perhaps, though, the scientist as law-maker is an over-rated idea. Isaac Newton himself was a member of parliament yet, according to legend, he spoke only once, to ask for a window to be closed. With anachronism, the British House of Lords is the parliamentary chamber strongest in scientists.

Karl Czernetz, Austrian member of parliament, was the most agitated speaker at a meeting in Vienna of European parliamentarians and scientists in 1964. Scientific programmes were of such importance to the life of the community that parliamentarians had to share in their preparation; yet Czernetz wondered whether the experts really thought such co-operation desirable. Some scientists, he believed, were opposed to partisan politics, especially concerning science; nor did Czernetz suppose it would be easy to attract scientists into government. 'Parliaments must be taken as they are and made to evolve.' He thought that parliamentarians should be willing to go back to school, as it were, in order to understand scientific problems. It was for politicians to be 'the engineers of society and the educators of humanity', and for politics to be 'an applied social science in harmony with ideals' (Czernetz, 1964).

Present reality is less imposing and some observers have always been sceptical, like the French mathematician, Henri Poincaré (1913). 'In England it is said that Parliament can do everything, except to change a man into a woman. It can do everything, I should say, except to deliver a competent judgement in scientific matters.'

'A sort of archaic fairyland, stripped of real power and useful only as a passive forum for the warring party machines,' was how Richard Crossman (1966) described his own legislative chamber. As leader of the British House of Commons, he was introducing a modest clutch of innovations which included the creation of a select committee on science and technology. Austen Albu, an engineer M.P., had pressed for such an innovation. He had lamented that British parliamentarians were 'singularly uninterested' in the hints the government gave in 1945 that it intended to develop nuclear weapons (Albu, 1964). The first British minister for science, Quintin Hogg (Hailsham, 1963), had contrived to write a book on science and politics without mentioning any role for parliament.

Anyone wanting greater involvement of parliament in debate of scientific and technological policy should be careful about what he has in mind. There are some striking object lessons from the US Congress, where legislators have insisted on making scientific policy themselves.

In the late 1950s, the Congress wished upon the National Institutes of Health money for finding a cure for cancer. The idea was simply to test many materials, on the off-chance that some of them would turn out to be useful in cancer chemotherapy. Years later, an independent committee under Dean Wooldridge (1965) was to report to the US president that, because the availability of money exceeded the availability of sound ideas, some unsatisfactory effects inevitably resulted. By then more than $30 million a year was being spent on the cancer chemotherapy programme; more than 25,000 agents had been tested on three million experimental animals. An appendix to the Wooldridge report commented: 'The resulting collaborative programmes reflected legitimate health needs recognised by the Congress, the NIH and the lay public. In general, however, their design did not allow them to exploit the serendipitous nature of basic research.'

Another case of a project falsely conceived and sustained only by the enthusiasm of the politicians was that of the nuclear powered aircraft. First thoughts showed heavy shielding would be needed to protect occupants from radiation from the reactor, so the aircraft would have to be enormous; furthermore, if such an aircraft crashed, there was no way of safeguarding the countryside against nuclear contamination. In spite of these objections the project was instituted in the late 1940s, and continued through the 1950s, until nearly a billion dollars had been spent merely to prove the correctness of these self-evident objections.

Sometimes, scientific matters become straight (or crooked?) issues of political influence. Here a classic case, again American, was the Mohole.

In 1951 the geophysicist Walter Munk proposed a scheme for drilling a hole in the ocean bed, right through the crust of the Earth, to where it met the main body, the mantle, at the Mohorovicic discontinuity (hence the name, Mohole). It was seen both as a worthwhile scientific undertaking and as a means of stimulating public and professional interest in the Earth sciences. The American Miscellaneous Society, affiliated to the National Academy of Sciences, was charged with developing the idea. The slogan was, 'The ocean's bottom is at least as important to us as the Moon's behind'. The first estimate was that the project would cost about five million dollars.

In 1961, the National Science Foundation invited companies to submit bids for drilling the hole. Brown and Root of Houston, Texas, was awarded the contract. Careful evaluation of the bids had shown two other companies to be suited for the task; there was comment on the fact that the president of Brown and Root was a close associate both of the vice-president, Lyndon Johnson, and of the Congressman, Albert Thomas, who was responsible for the NSF budget in the House of Representatives. By this time the cost was ten times greater than the original estimate and Joe Evins, a member of Thomas's subcommittee, objected to the expense.

A site for the deep hole was provisionally selected near Hawaii, but costs rose further and Earth scientists turned out to be far from unanimous in support of the project. In 1966 Evins became chairman of the subcommittee and blocked further funds for Mohole. Johnson, by then president, appealed to the Senate to save the project. The estimate had risen to $125,000,000. The Senate was willing to comply—but it emerged that the chairman of Brown and Root had made a substantial contribution to Democratic Party funds a few days after Johnson's appeal to the Senate. No connection was proved, but Mohole finally dissolved in an aura of political scandal. Science policymakers were abashed. (For a fuller account of the Mohole affair, see Greenberg, 1967.)

LEGISLATORS' DIALECTIC

The stories of legislative folly notwithstanding, the US Congress led the world's parliaments in taking interest and gaining influence in technical policy. It benefited in several ways: from the existence of a diversity of specialized committees of congressmen and senators able to hold quite detailed technical hearings; from the presence on the staff of these committees of scientifically trained people; from a Legislative Reference Service employing scientific staff and providing detailed information and analyses; and from a close liaison with the National Academy of Sciences. The American Association for the Advancement of Science and the Brookings Institution provided classes on scientific topics for congressmen and senators. More generally, the men on Capitol Hill had the advantage over most other parliamentarians in their office and staff facilities.

A Science Policy Research Division was created by the Library of Congress in August 1964, to match the new science advisory facilities of the Executive. In its first nine months it dealt with 350 inquiries from senators, congressmen and subcommittees. It prepared forty-one analytical studies including two massive reports on government weather programmes and on fifteen years' work of the National Science Foundation.

Congressional committees could make research workers and engineers squirm by their questioning, and mercilessly scrutinize budgets for science and technology. But could the members make themselves well enough informed? While testifying before the Senate subcommittee on government research, the political scientist John Plank (1966) turned the tables on his inquisitors by asking the chairman, Fred Harris, about the willingness of members of Congress to think constructively—in this case, about the social sciences.

Harris replied that reading time and reflecting time were at a tremendous premium. 'I thought when I came to the Senate of the United States that it would be a wonderful place to think new thoughts and learn new facts, and I am more and more impelled to the view, impractical as it is . . . that we need a sabbatical year during our six-year term. . . . Unless a member of Congress has to make a decision, he probably doesn't have much time to go outside of what comes immediately before him.' For this reason one should not be too easily impressed by those expert staffs for the congressional committees, or that high-powered Library Reference Service. Learning requires more than a little effort.

Besides his primary responsibility as a watchdog of the people over the actions of governments, the parliamentarian's role in technical policymaking is difficult and questionable. As a general-purpose citizen, a parliamentarian is not likely to be successful in making detailed policy. Nor is it clear that he should try; if a committee of members of parliament proposes certain policies which are adopted by the government, then the right of other members of parliament to criticize or oppose the policies may to some extent be compromised.

Practical questions of this kind are, however, superficial compared with the deep-rooted political malaise of the scientifically advanced nations. In an historical study, Ronald Butt (1967) found no evidence to support the widespread belief, illustrated by Crossman's remarks quoted earlier, that power and influence in the British parliament had declined in relation to the executive gains. Butt argued that the basic purposes of parliament remained those of preserving the legitimacy of government, maintaining the freedom of the citizen, ensuring that the dialectic of politics never ceases, channelling into 'broad and manageable streams' the choices available to the community and bringing the government into constant consultation with elected representatives. Butt's conclusion was that difficulty arose, not so much in the processes of parliament, as in the existing party structure; political issues lay increasingly across party lines.

The conventional ideology of political parties is uninstructive about the merits and drawbacks of particular technological developments. If science and technology are 'above politics', as some affirm, there is no reason to expect much interest or enthusiasm from parliamentarians, who can then partake in science policymaking only as supernumerary civil servants.

A NEW PRIESTHOOD?

The loss of political control over the course of our technological civilization is, for many, simply the fault of 'the scientists'. In some kind of professional conspiracy they are supposed to be forcing on the rest of us all the wildest possibilities arising from their experiments. They do so either in blind indifference to the consequences or in deliberate quest of power, operating behind the scenes in government.

So much surface plausibility encases this popular myth that it has to be taken seriously. It is true, for example, that experts press enthusiastically for their pet projects. Furthermore, leading research workers and technologists are often privy to operations of 'closed politics', where important decisions are made, shaping the world of the future. Yet almost never, except in matters of scientific judgement on the tactics of research, can 'the scientists' make important decisions about government policy without approval of their political bosses.

Well-informed observers have endorsed the idea that 'the scientists' form a new priesthood. Don Price (1966), a political scientist of Harvard, said that scientists had replaced the clergy as the most important intellectual force to be reckoned with in human affairs. 'New physicist is but old presbyter writ large,' he remarked, modifying Milton.

The confrontation of scientists and clergy over birth control, and the gradual victory of the scientists, provides a nice direct test of this view. But we hesitate to follow Price when he goes on to liken scientific institutions to churches. Besides the fair analogy between scientist and priest, that between science and a priesthood looks weak and misleading. Ralph Lapp (1965) observed: 'there is no conspiracy of scientists about to seize the reins of government . . . they show no signs of aspiring to political power'.

Religious priesthoods, on the other hand, have very often sought political power. A characteristic of a priesthood is its creed, embodying a view of life and therefore of society. Except in their adherence to the internal rules of research, which have little to do with practical applications of science, scientists as a group have no common creed. Nor have they any sense of social or transcendental purpose beyond the day-to-day enlargement of knowledge and *ad hoc* responses to technical problems and opportunities. Least of all do they have a programme—a coherent plan for creating a particular kind of world by the development and application of particular technologies. It might be better if they had: at least the rest of us could then know what it was, and decide whether we liked it or not, in whole or in part. As things stand the scientists are fragmented in specialist groups and not one of them knows where he, or science, or the world is going or should go just now. What price priesthood?

The British astrophysicist Fred Hoyle (1968) asked: 'Why in fact do we do physics?' and supplied his own answer: 'The real motive, of course, is a religious one. . . . It seems to me that our only chance for the future is to recognize ourselves for what we are—the priests of a not very popular religion—and to see what we can do about it. If society will not accept us for what we are we must change society.'

That fighting talk does not yet represent a common viewpoint among scientists. Even among those most aggressive for scientists' rights, the policy has been to play the game society's way, and base their claims on the utility of science; Hoyle said, quite rightly, that this attitude was basically dishonest. There is really no sign that the professors are about to marshal the militant students and march on the capital. Yet attitudes among scientists might change quickly, either because of professional frustration or because they see society about to hit disaster.

THE CASSANDRAS

In the late summer of 1962 an earthquake in western Persia caused many thousands of deaths. Immediately before, a party of Soviet seismologists, hundreds of miles away in the Caucasus, had exploded two-and-a-half tons of high explosive so that they might study the waves from a little artificial 'earthquake'—an entirely commonplace type of experiment. The idea that the two events were linked was absurd, yet it was openly discussed in press and radio comments. It was symptomatic of public concern about the activities of research workers and technologists interfering with nature—like the folklore then current that nuclear bomb tests affected the weather.

At that time, before alarm became much more general, we ran a leading article in *New Scientist*, entitled 'Science in Disrepute', pointing out that such alarm, though founded on ignorance, was not allayed by some of the actions of scientists. It instanced the admission just then made by the US Defense Department that a megaton bomb exploded at high altitude had created a more intense, wider and longer-lasting radiation belt around the Earth than had been expected—bearing out forebodings of foreign physicists. Other blunders recapitulated included the injury by fall-out to the Japanese fishermen in *Lucky Dragon* and an accidental death from pneumonic plague at the British biological warfare centre at Porton. On the civil side, the leader noted the thalidomide drug disaster and the consequences of the indiscriminate use of antibiotics and pesticides. And it concluded: 'Recent events have sounded a warning that the scientific community will ignore only at great risk to the prestige—and consequent tolerance and support—which it at present enjoys.' (Calder, 1962)

By the mid-1960s, however, such sounding of tocsins had become too loud and repetitive, especially in the United States. At the end of 1964, a committee under Barry Commoner, a biologist of St Louis, presented a report on 'The Integrity of Science' to the American Association for the Advancement of Science, which cited a number of events that gave cause for concern about the erosion of scientific principles. Ralph Lapp (1964) concluded his book, *The New Priesthood*, with a chapter on 'The Tyranny of Technology'. Later, Commoner (1966) published *Science and Survival*, in which he said that technology was getting out of hand. A subcommittee of the

---

I. TWELVE STANDARD HORRORS

Thalidomide disaster: malformed babies follow use of new drug (1)

Abuse of antibiotics: resistant strains of bacteria appear (1, 3)

Abuse of pesticides: killing of wild life and environmental contamination (1, 3, 4)

Lucky Dragon incident: fishermen caught in unexpectedly far-reaching H-bomb fall-out, one dies (1, 3)

Starfish high-altitude H-bomb: 'wrecking' of Earth's radiation belts despite protests of astronomers (1, 2, 4)

Water pollution: especially (at one time) from detergents not susceptible to natural break-down (2, 3, 4, 5)

Air pollution: especially from power stations and automobile exhaust (3, 4, 5)

Modification of atmosphere: by carbon dioxide from fuel, and by exhausts of jets and rockets, with unknown effects on the Earth's climate (4, 5)

Over-fishing: technological aids cause depletion of fishing-grounds, near extinction of whales (5)

Bugging devices: electronic invasion of privacy (5)

Sonic boom of supersonic airliners: pending (3, 4)

The great power black-out: north-eastern USA, 9 November 1965, a failure of automatic systems (4)

Cited in sources as indicated: (1) *New Scientist* editorial (1962); (2) AAAS *The Integrity of Science* (1964); (3) R. Lapp *The New Priesthood* (1965); (4) B. Commoner *Science and Survival* (1966); (5) Daddario subcommittee report (1966).

US Congress, under the energetic, science-minded Emilio Daddario, chimed in with anxious paragraphs on 'Dangers of New Technology'.

Table I, on the preceding page, itemizes 'standard horrors' raised in these various publications. The table is not exhaustive —one might include the accidents with nuclear bombs in Spain and Greenland, and the 'chocolate mousse' of crude oil from *Torrey Canyon* and from the off-shore field at Santa Barbara. But much anxiety in the 1960s focused on these points. Most are scandals of one kind or another but they are not outside the scope of human control and they should properly be balanced with a list of some beneficial consequences of research and technology during the recent past (Table II, below).

The lesson of Table I is that we must be much more alert to unlooked-for side-effects of innovations; there must be better anticipation of side-effects; also more and earlier discussion of them. A great need for humility and a readiness to admit to ignorance, are essential when research workers and technologists are called upon to advise about innovations. A couple of tons of high explosive cannot cause an earthquake, but it turns out that building dams to store water can do so.

In the 1950s and 1960s a growing volume of research and technology had to be applied to 'remedial science', putting right things that were going wrong with earlier applications: to monitor radioactive fall-out and other forms of pollution; to make new poisons or drugs against insects or microbes that had evolved resistance to earlier chemicals; to develop artificial limbs for thalidomide children, and so on. And misgivings grew as the image of science was tarnished. Lyndon Johnson (1968) warned researchers: 'An aggrieved public does not draw the fine line between "good" science and "bad" technology.'

Daddario's congressional subcommittee proposed a 'Technology Assessment Board'—Daddario thought the acronym TAB was appropriate because it would keep a tab on possible dangerous side-effects inherent in new technology and would provide a public 'early warning system'. Daddario subsequently developed the idea in a bill he presented to Congress for the creation of an independent Technology Assessment Board; although he expressed some doubt about the conception of the Board as a practical means of achieving his end, he thought the proposal would stimulate discussion on how to tackle the problem. In the US Senate, John McClellan of Arkansas also sponsored a Bill with similar objectives.

The boldest American proposal for the control of technology was that of Wilbur Ferry (1967) of the Center for the Study of Democratic Institutions, who proposed rewriting the US Constitution to limit technology, 'when it can no longer help but

---

II. TWELVE RECENT BOONS

Nuclear power: relieves the human species of fears of rapid exhaustion of energy sources.

Microbiology: infectious disease can now be largely eliminated.

Space technology: much improved weather observation and telecommunications.

Desalination of water: makes deserts habitable, perhaps one day cultivable.

Computer: eliminates much clerical drudgery and aids in mastery of complex systems.

Plant and animal breeding: vastly increased yields of food per acre.

Technology in general: rising productivity means rising living standards, increased leisure.

Technology in general: machinery, detergents, synthetic fibres, etc., greatly reduce domestic drudgery.

Technology in general: improved paints, dyes, materials and designs make the man-made world more attractive and colourful.

Air transport: safer, faster, cheaper.

Television: medium of entertainment and instruction of unprecedented potency.

Psychiatry: much mental illness now susceptible to therapy.

---

only harm the human condition . . . I can conceive of no way of effectively regulating and dispersing the power of technology other than by placing it under the Constitution.'

But the Cassandras were growing shrill, and obsession about side-effects carried two dangers. First, it could easily frighten people out of their wits, in the sense of losing the wit to think clearly about the benefits and risks of technology. In the words of a seventeenth-century nursery rhyme:

> *'If there had been no projects,*
> *Nor none that did great wrongs.'*

Now look at what is labelled as a table of non-standard horrors (Table III). Its contents are of a different order of importance from most of the standard horrors. They are non-standard, not in the sense that few people are concerned about them, but because these are not the issues most commonly cited when concern is expressed about the use and abuse of technology.

Why not? Simply *because* they are of a different order of importance, so that they rise out of the area of cool, administrative politics where issues about research and technology are commonly supposed to dwell, into the zone of hot, ideological politics?

It is one thing to express concern about the arithmetical error that killed *Lucky Dragon*'s radio operator; quite another to organize the dismantling of nuclear weapons systems. To seek legislation against noxious exhaust from automobiles takes less boldness than to circumscribe the freedom of the driver to go where he pleases and drive as he likes—in a country where most voters are drivers. Again, anyone must grieve for the thalidomide children of Europe and see that they are cared for; it requires deeper thought to wonder why those thalidomide mothers needed sleeping pills, and a far sterner imagination to make sure that love and pharmaceuticals are deployed on behalf of malnourished, parasite-ridden children in Africa.

In other words, of greater consequence than the side-effects of technology are the central effects and direct uses of technology. Mishaps and inefficiency are matters for attention, but not more so than motives and morals.

III. TWELVE NON-STANDARD HORRORS

Nuclear weapons systems: half the world lives in continuous mortal danger; weapons stockpiles growing and spreading to new nations.

Biological weapons systems: under active development, although illegal by international law; threaten to outclass the H-bomb.

Space technology: gives the super-powers potential domination of the planet by communications control and orbital surveillance; arms race of space weapons.

Ocean technology: promises efforts by powerful nations to grab the resources of the oceans; arms race of ocean weapons.

Computerized intelligence: world politics already under centralized surveillance.

Agricultural and food technology: with it, people in the rich countries die because they eat too much; without it, people in the poor countries die because they eat too little.

Technology in general: makes the rich richer and the poor at least relatively poorer; prospect of alliances of white, rich northern nations versus poor, coloured southern nations.

Technology in general: destructive of natural habitats and threatening remaining enclaves of wild life.

Automobile transport: costly, unsafe, noisy, poisonous, destructive of cities and countryside.

Television: vehicle in many countries for advertiser's pap or government propaganda.

Psychopharmacology: threatens the sanity of a generation.

Medical technology: continuing emphasis on ever more costly therapy, rather than prevention.

POLITICAL LOADING

Medina Sidonia was dismayed when dawn broke on the last day of July 1588. He was off the Eddystone with his Armada, the mightiest fleet ever assembled, afloat upon the Enterprise of England for bringing that Protestant country back into the Catholic fold. The English fleet, which he had spotted far ahead to the east the evening before, had somehow worked round to windward behind him during the night and had seized the tactical advantage. Howard of Effingham's fighting ships were as unorthodox as Elizabeth's religion. By John Hawkins' design they were slimmer, faster and more weatherly than any built before; William Wynter had armed them with long-barrelled culverins of high muzzle velocity, in place of fat cannons of short range. These were the ships that for two weeks were to hustle the Spanish galleons through the Channel and as far as Scotland, there to leave them holed and splintered to face the northern storms.

This confrontation took place when modern science was only embryonic in a few men's minds, but the decisive technological innovations flowed from a similar source. Escape from spiritual dogma encouraged adaptability of other kinds, which helped the Protestants to keep their island from the Catholics and later from other leaders with potent creeds. Medina Sidonia's shock was to be precisely matched by that of Hermann Goering's pilots in 1940, who found the British fighters coming at them from above and out of the sun, put at that advantage by word from the radar stations.

In philosophy—natural, moral and political—the islanders turned their back on Europe and its weighty dogmas. As if by design, they drove their non-conformist brethren to carry the seed to North America. The concept of the 'Anglo-Saxons' is ethnic nonsense, yet English-speaking empiricism, stiffened with puritanism and technology, overpowered the world. And as the British sent the last of their battleships to the breaker's yard, the Americans built their first Polaris submarines.

In the 1960s, British governments turned peaceably to Europe for the first time in four centuries, and the dowry that Harold Wilson offered in despite of Charles de Gaulle was not empiricism, nor puritanism, but technology. By then, in styles of science, the English Channel seemed wider than the Atlantic.

The beginning and practice of systematic research cannot be separated from moral and political ideas. The men who founded the Royal Society of London were strongly Puritan in their outlook. They explicitly dedicated their work to the glory of God and to social welfare. The same religious view of the individual answerable directly to God, which promoted non-conformism and diligence in worldly affairs, sanctioned the experimental approach to nature. The American sociologist, Robert Merton (1962), found evidence from diverse sources, including Catholic ones, confirming that Protestants were much more likely to become distinguished scientists than Catholics were. This conclusion, that success in research depends on general attitudes, is borne out by impressions of the state of science in various countries and at various times, especially by its failure to flower so far in this century in the Catholic countries of Southern Europe and Latin America, and by the lack of originality in Communist research and technology.

The origin of present attitudes to the uses of knowledge lies only partly in the acquisition of powerful science. Religious, social and philosophical ideas, evolving over many centuries, gradually made men abandon the idea that nature was itself divine, in favour of the idea that men should collaborate with God in changing the environment. Clarence Glacken (1967), geographer at Berkeley, showed how conservationists had to battle even in mediaeval times, while Lynn White (1966) of Los Angeles laid blame for the ruination of the global environment squarely on the Christian axiom that nature had no reason for existence save to serve men. It is a short step to believing that change itself is holy.

Today, the purposes and policies for research are caught up by the world-wide political pressures towards mass education and economic growth. No area of public life, from the church and the sports field to the stock exchange and the law courts, can now escape the impact of science and technology. The concepts of private capitalism and Marxist socialism both began to look extremely tattered in the rich countries in the 1960s.

International affairs are now dominated by the existence of weapons of incredibly destructive power, created with the aid of science; by the competition in technological exports; and by the effect of technology in widening the gulf between rich and poor, white and coloured. Looming up are black clouds of international dispute about the uses of space satellites, the ownership of the ocean and the integrity of the atmosphere. Science and technology are filling the foreign minister's In-tray.

They also invade every other department of government, and not merely those traditionally concerned with the hardware of defence, industry, transport or communications, or with the technologies of agriculture. The ministry of education is now confronted with demands for more young scientists and engineers, with the problem of teaching youngsters who will inhabit a very different world from ours, and with developments of computers and other teaching aids that cannot be ignored. Administrators of justice also have the computer to assist or perplex them, together with a multiplicity of new techniques for crime and the detection of crime, which in turn raise basic questions about the prospect of privacy versus Big Brother. While the minister of labour frets variously about shortages of skilled workers or unemployment caused by automation, the minister of health sees advances in medical and surgical science that could empty the treasury if they were fully applied for the well-being of the citizens. The pattern of political life changes visibly with the use of television for propaganda and debate; invisibly with the computerization of government and the advent of allegedly scientific techniques of policymaking. The organization of research and technology themselves has become a new pre-occupation of government.

In spite of all this, science is commonly regarded as falling outside the area of controversial politics. We inhabit Technopolis, a society not only shaped but continuously modified in drastic ways by scientific and technical novelty, yet even for administrative purposes science is still treated, in most countries, as peripheral to the traditional pre-occupations of politics. In the USA, USSR, UK and Sweden, where the importance of research and technology are well recognized, argument about them is far more likely to be concerned with efficiency than with morals, with speed rather than with direction.

Nevertheless, attitudes to science policy are evolving very rapidly. In the late 1960s, more and more commentators and the first few politicians were beginning to realize that advance in science and technology forced a re-examination, not just of how best to achieve pre-existing social goals, but of what the social goals ought to be. With only mild exaggeration, forgivable in one who'd been round the Moon and back as skipper of *Apollo 8*, Frank Borman (1969) testified: 'Man can now do anything he wants to, technically.' The question remains unanswered —what?

Besides the different styles of science and its management in different countries, there are wide disparities in the sophistication of science policymaking. Some 'advanced' countries have scarcely reached the starting post, where you decide that research is worth patronizing on a significant scale. Even the leading scientific nations are just at the point of learning how to encourage technological innovation in all economic and administrative activities, and how to look ahead to anticipate the consequences, for good or ill, of those innovations. The organization of serious public debate about the uses of science has scarcely begun. Still less is there any ordering of moral and political attitudes that might give some coherence to the argument, or a pattern for new policies. I aim to identify some general and particular political issues of our scientific age; also ways of reconciling long-range planning with democratic debate; and, for the debate, an order of battle.

CHAPTER 2

Patrons and Purposes

EVEN before the invention of the telescope, the Danish observatories of Tycho Brahe, the first modern astronomer, cost Frederick II the equivalent of nearly $2 million to equip and run. His successor gave up this expensive hobby and let Tycho go down the brain drain to Prague (Gerholm, 1967). As a result, the Holy Roman Emperor Rudolph II won the credit as the patron for the first big step in modern physical science. Sound experimental philosophy was emerging from the mists of astrology and alchemy, as the sixteenth century ended and the seventeenth began, when one of the most fruitful collaborations in the history of science occurred in Rudolph's court in Prague.

There, in the baroque fortress on the hill, were foregathered men of a Faust-like mixture of devilry, wishful thinking and great learning. Neither they nor their imperial patron had any doubts about the utility of their enterprise. They sought to predict the future from the planetary motions; also to turn base metals into gold, to make cunning automata and to find the means to everlasting life. In the White Tower, you might find imprisoned a visiting alchemist suspected of concealing the secret of goldmaking, while a more successful conjurer was being honoured by the emperor as a true follower of Hermes.

You would also find, in the year 1600, Tycho Brahe and his young refugee assistant, Johannes Kepler, puzzled about the orbit of the planet Mars. Tycho died in the following year but after several years of further calculation from his collaborator's measurements, Kepler discovered that the orbit of Mars around the Sun was, contrary to all previous theory, an ellipse. From Kepler's results Isaac Newton was to derive his law of universal gravitation. In dedicating his report to Rudolph, Kepler wrote:

'Finally the enemy [Mars] resigned himself to peace and by the intervention of his mother, Nature, he sent me note of his defeat, became a prisoner on parole, and Arithmetic and Geometry escorted him unresisting into our camp.'

(Taton 1965)

He followed this arch account of his discovery with an outline of plans for research on the orbits of other planets, and an application for money. With its first breaths, infant science hollered for funds.

Then was the world truly vexed with astronomy, not least the Church of Rome, when Galileo began seeing impossibilities through his telescope. The stargazers needed expensive instruments, computations and expeditions, like James Cook's voyage to Tahiti to observe the transit of Venus. Kepler dreamed of manned flight to the Moon; three and a half centuries later, when most astronomers' thoughts had strayed far beyond the solar system, engineers put the project in hand at astronomical cost. They set out also to capture Mars and Venus, not with Arithmetic and Geometry, but with television close-up and parachuted capsule.

Over the centuries, science found other patrons and purposes. Tsar Peter the Great of Russia embodied in his person the possibility of a state founded on science, technology and absolute political power. Anyone who nurses, today, the idea of a benign scientific dictatorship might hesitate about taking this young savage giant as his model. How Peter travelled in Western Europe, sought Dutch know-how in shipbuilding and printing, visited the botanists, anatomists and astronomers of the Netherlands and England and, by ruthless reforms, opened the window of his homeland to the advanced ideas of his time—this story is well known, and the history of modern Russia begins with him. He organized geodetic and mineral surveys of his huge country, often involving journeys of great daring, and these were the finest scientific achievements of his reign.

Peter lived long enough to order the establishment of the

St Petersburg Academy of Sciences at his new capital on the Neva, though not long enough to see it come into being. He was himself a member of the French Academy of Sciences, and he conceived his own Academy over years of study, influenced by his correspondence with the mathematician Gottfried Wilhelm Leibniz, founder of the Berlin Academy of Sciences. Three durable traits of science in Russia originated in Leibniz's influence and Peter's actions. One was the inclusion within the sciences of social, historical and philosophical studies, as in the German *Wissenschaft*. Another was the close identification of science with the political needs and aspirations of the Russian state. A third was the emphasis on mathematics, a field in which Russian research workers have shone ever since. A Scotsman, Henry Farquharson, was largely responsible for introducing Western mathematics into Russia during Peter's reign, in the Mathematical and Navigational School, where he taught seamen as well as students.

On paper, Peter launched Russian science in splendid fashion. But it faltered, almost from the outset, for want of indigenous scholars—the watermill lacked water—and because of the stifling influences of an absolute monarchy. Teachers were underpaid and Mikhail Lomonosov, a heavyweight personality but a lightweight scientist, dominated the scene in the mid-eighteenth century. Catherine the Great read the enlightened French philosophers and sought to expand education, but finished her reign by banning the import of foreign books.

What must be counted a fourth durable trait in academic Russia was reported on one occasion by Leonhard Euler, the brilliant recruit to St Petersburg from the Bernoullis' school at Basel, who worked in Tsarist Russia as a mathematician for many years. When chided by the Queen Mother of Prussia for his reserve in answering her friendly questions, Euler replied: 'Madam, I have just come from a country where people are hanged if they talk.' (Vucinich, 1963)

Consider next, in the year 1790, the secretary of state of the USA in his office, carrying out scientific tests on a salt-water still—as part of his official duties. That Thomas Jefferson, as a senior cabinet officer, was better qualified than any other employee of the government to do the work lent plausibility to the activity. But he was not amused, the less so as the mixture alleged to assist in the desalination of sea water proved ineffective. His job was diplomacy and the responsibility for granting patents, casually laid upon him by Congress, was quickly becoming oppressive. When he complained, Congress responded by making patents automatic on application, without any vetting of the inventions.

Rarely were political authority and amateur enthusiasm for research so mixed as in the person of Jefferson. This was the man who had shipped the bones of a moose from New Hampshire to Paris, to confound the biological theories of the great Georges Buffon, according to whom American animals were necessarily degenerate and smaller than their Old World counterparts. Before he became President of the United States, in 1801, Jefferson had taken an active part in starting the US mint, but he failed to persuade Congress to adopt a decimal system of weights and measures; so to this day Americans use what are basically the ancient English units.

Jefferson's ideological dedication to limiting the power of central government scarcely encouraged federal scientific enterprise. Yet, as president, his major scientific coup directly served a national purpose. He sent two army officers, Merriwether Lewis and William Clark, on a scientific journey across the continent to the Pacific Ocean (1804–6). Jefferson himself laid down the technical objectives of the expeditions. Lewis took a crash course in celestial observations, botany, zoology, and anthropology, at the American Philosophical Society in Philadelphia. Jefferson assured the French and Spaniards that the aims were purely scientific; he told Congress, in requesting funds, that the object was to extend trading opportunities; his real purpose was to enlarge his nation's empire. Before the expedition left, Jefferson's ambassador bought the huge Louisiana territory from Napoleon for $12 million, but Lewis and Clark went beyond the Rockies and built a fort on the Pacific shore.

Their expedition was a success, but there were few American experts to study the specimens and information they brought back.

That the founders of the United States included men like Jefferson and Benjamin Franklin was no guarantee of effective policies for research and technology in the young nation. The Constitution implied activity on several scientific fronts, including census-taking and the granting of patents, but federal action in support of research was slow to develop—even in the creation of 'lighthouses of the sky', as John Quincy Adams called the observatories. The removal of the capital from Philadelphia to a sea of mud in Washington, in 1800, took the government away from an embryonic scientific community. An extremely cautious attitude of congressmen to any national scientific enterprise was to persist well into the nineteenth century. When James Smithson, an obscure British chemist and illegitimate son of the Duke of Northumberland, left a fortune of $500,000 to the United States for founding a scientific institution, Congress dithered for sixteen years before finally authorizing the creation of the Smithsonian Institution (Dupree, 1957).

DAYS OF FEAR

Reflexes were faster in 1957. The surprising thing about the launching of *Sputnik I* is that Americans were so surprised. For some reason the learned advisers, the intelligence services, and the academic students of Soviet policy had all failed to anticipate the event and world-wide reactions to it. Nikita Khrushchev had not even sought to treat it as a secret programme; on the contrary several months earlier his physicists had explicitly told the central planning committee of the International Geophysical Year—a world-wide collaborative research enterprise—of the intention to launch Earth satellites during the period July 1957 to December 1958. They even gave rough indications of the launching site and of the orientation of the intended orbit. The unimaginative expertise of the Washington establishment led it to neglect the likely interest in the first

Earth satellite, so that, all unprepared, John Doe woke that morning to find that his nation was not, as he had supposed, master of the universe.

Now that hundreds of satellites have been launched, it is hard to recall the impact of that first clumsy orbiting vehicle. Not only was it a sensation but the effects were sensational and will be felt for decades to come. Because of the impetus to research, technology and education, from the shock of *Sputnik I*, Western countries benefited in a way that Moscow never could have intended. Many believed that the free passage over American territory of the beeping satellite represented a major shift in the military balance of power to the Soviet Union. The rest of the world thought that the Soviet Union was beginning to eclipse the United States in that supreme product of Western capitalism—advanced technology. The ponderous spacecraft, circling the world every hour and a half, seemed to many to turn it upside down.

If the *Sputnik* did not exist, the Americans would have had to invent it; something was needed to save research and technology from being stunted for lack of further public funds. They had been shooting up rapidly. Part of the growth, especially in the more expensive items, stemmed from military interests in rapidly advancing technologies—notably those of nuclear energy, electronics, aircraft, guided missiles and ballistic rockets. The erosion of the Americans' 1945 advantage, as the Russians developed their own A-bombs, H-bombs and delivery systems, had only made the Americans bustle more. But another part of the growth was civilian and more spontaneous. Research workers had come back from the war to find big ideas waiting to be taken up—in solid-state physics as well as in nuclear physics, in biology as well as astronomy. In the universities, more and more young Americans came forward as freshmen and, as the faculty staffs and their graduate students grew in numbers, so did the scope and costs of their research. Tremendous commercial opportunities, too, had quickly become apparent, in television, computers, drugs and dozens of other new items, from long-playing records to unconventional metals and alloys. On such things were great industries founded and much of our present material environment created.

A barrier remained. For the general public, for the slower businessmen, and for Congress, research was an activity of odd-balls. Its connection with the citizen's prime duty to make profits, or even with outsmarting the Communists, remained obscure. Everyone could see that nuclear bombs and missiles had something to do with research—but if that was all, why not just hire the necessary men, put them under military guard to stop them selling the secrets to the other side, and let them get on with it? Charles Wilson, Dwight Eisenhower's defence secretary, made wisecracks about research that told why fried potatoes were brown and grass was green, and not long before *Sputnik I* he abolished the post of his assistant secretary for research and development (Lear, 1957). But the natural growth of research—with ideas breeding discoveries breeding new ideas, and with young men crowding into the lecture rooms and the laboratories—had brought it to a point where the budgets were quite big enough to catch the chilly attention of legislators and company accountants. If they were to go on growing, there had to be a change of attitude. That the *Sputnik* gratuitously achieved. If it was by research and technology that Khrushchev meant to bury capitalism, America had the answers—and that went for fried potatoes, and grass too. The post-*Sputnik* USA was marked, as one commentator put it, by 'an unlikely coalition of liberal advocates dedicated to social improvement and militant patriots devoted to the defeat of communism.' (Lekachman, 1967)

Eisenhower had to go on television after *Sputnik II* went up with a dog aboard, to reassure his countrymen. He appointed the president of the Massachusetts Institute of Technology, James Killian, as his science adviser—a new post. Thereafter, research and development for space sharply increased, and the civilian National Aeronautics and Space Administration came into being. Congress took the closest interest in the details of the space race, and every success and failure of Russians or Americans reverberated on Capitol Hill. The Senate created a select committee on space activities and its chairman was an ambitious senator from Texas, Lyndon B. Johnson.

Purists grieved to see so much nonsense talked about the *Sputniks*. This Soviet 'first' in a single field could not erase a clear American advantage in almost every other branch of research and technology. The American satellite programme had been delayed, not by technical incompetence but by inter-service rivalries compounded with general lack of political interest. In any case, it was distressing to see how 'science' came to be equated with the technology for hurling packages into space. But the less squeamish among the research leaders seized the opportunity with both hands, to batter open the treasury doors. They waved phoney statistics about Soviet research and education to confuse the guards. And they re-emerged laden with gold for research and education that had nothing to do with outer space.

For a start, the budget of the National Science Foundation was trebled and science education received a massive injection of federal money under a law significantly entitled the National Defense Education Act. After that, support continued to grow for nearly a decade, until the Vietnam War became expensive. Three years after *Sputnik I*, Eisenhower (1961) was shaking his head over the risks to public policy of the new order: 'Yet, in holding scientific research and discovery in respect, as we should, we must also be alert to the equal and opposite danger that public policy could itself become the captive of a scientific-technological elite.' Research workers had won their gold, but from now on the politicians would be at least wary of them.

FORMEZ VOS BATAILLONS!

That grand philosophy gets you nowhere may be the supreme lesson that France teaches the world. A succession of brilliant French philosophers failed in the search for an ideal political system and for a consistent and humane set of attitudes for confronting the modern world. Intellectual factions waxed and waned; the teacher whose influence remained supreme was

Napoleon Bonaparte. Styles of centralism date back to Colbert, but the spirit of pragmatic nationalism, and the educational system that embodied it, came from the Emperor.

Between the Pantheon and the Seine, where rue Descartes runs across the Montagne Sainte-Geneviève, is the buff stone gateway of the École Polytechnique. The entrances to the site are guarded by troops, for this is an arm of the state as well as an institute of learning; it is commanded by a general. Napoleon called the École Polytechnique the goose that laid his golden eggs. Ever since, it has dedicated technical knowledge to the glory of France. Its pupils, known as the 'Xs' because of the crossed cannons of their emblem, are selected by fierce competition and educated intensively in mathematics and the physical sciences. On the basis of their results they go to the officer corps of the armed forces or to technical departments of the civil service. They have great opportunities beyond, not only in the public service but in research, industry or politics, and remain a most potent influence in French public life.

Yet the ascendancy of the *grandes écoles* had a more subtle effect on French research and government. The best students, *crême de la crême*, by going to these establishments, deprived the universities of their talents. Moreover, in spite of their book-learning in technical subjects, their subsequent public service usually prevented them from engaging in research. The Anglo-American tradition of having first-class graduate students blow glass, solder resistors or, in wading boots, collect biological specimens, was lacking for the best young minds of France. The white collar, the office desk, a reluctance to dirty one's hands—such was the conditioning of the typical products of the *grandes écoles*. Research for them was not something beloved for itself, but a tool for great enterprises.

The milieu of order and reason that Napoleon created was like a battleship, formidable, inflexible and obsolescent. In the practical pursuit of new discoveries and inventions, performance was only middling good. From time to time, the state-run educational and scientific establishment was refitted to suit changing times, as was happening in the 1960s. But the system had a tendency to freeze again. The molecular geneticist Jacques Monod (1963) remarked: 'Since Napoleon there has been no university in France in the proper sense of the term.'

While every government was bending research and technology to serve its military and economic purposes, and its prestige, that in Paris was franker and more resolute about it than most. 'Nowadays a scientific research policy is indispensable to a country's independence,' declared Charles de Gaulle's government. The drums of the Fifth Republic beat for research and technology.

And the *Marseillaise* resounded as the fat hull of the submarine *Le Redoutable* went down the slipway at Cherbourg one March day in 1967. The President of France launched her himself in part fulfilment of a dream that had haunted him and exercised many of the researchers and engineers of his country since his return to power nine years earlier. The dream was of national independence founded upon a do-it-yourself nuclear second-strike force—meaning, in the jargon of the deterrent theorists, the capacity for posthumous reprisal after the destruction of the homeland. French submarines powered by French nuclear reactors were to carry French missiles tipped with French H-bombs. *Le Redoutable* was the first boat in a programme that mimicked, even in the number of missiles she carried, the Polaris system of the US Navy. Her cargo would be capable of smashing sixteen cities by blast, heat and radiation.

Before de Gaulle came back to power, work had been going on for six years on the A-bomb. To inquirers asking how the bombs could be delivered to their targets, the answer was 'We'll have to send them by mail'. The 60-kiloton A-bomb used, as its explosive, plutonium made from natural uranium in nuclear reactors at Cadarache. To make the core of the H-bomb, and provide fuel for the nuclear submarine, meant building a very elaborate and expensive gas-diffusion plant for separating from natural uranium the relatively light and rare explosive component, uranium-235. That was duly done at Pierrelatte. While a small supersonic bomber, Mirage IV, was conjured into existence to carry the A-bomb, the protracted business began of developing ballistic missiles for the 1970s, for launching from the ground and from the submarines.

Thus French nuclear and rocket engineers were to provide the supposed proof of national dignity in the modern world: the power to erase cities from afar. Out of the Precious Stones series of rockets also came the Diamant satellite launcher and its successors, to provide another symbol of modern technological nationhood as France became the third nation able to put a satellite in orbit around the earth. There was a firm intention to launch communications satellites; a scheme for telephone links via satellite to the French-speaking ex-colonies and Quebec showed awareness of the political importance of this means of world-wide communications. And as communications satellites are best launched from a site near the equator, especially if rockets are not very powerful, the plan included construction of a launching range in Guiana, to replace the Hammaguir range in the Sahara.

Nor did the planners in Paris neglect 'inner space', as the oceans came to be known. With the Americans engaged in an expensive ocean programme, it represented yet another area for big endeavours by rival nations. As a leader and publicist of undersea exploration, Jacques-Yves Cousteau had done as much as anyone to alert his fellow men to the possibilities of the oceans, and the Fifth Plan gave Cousteau and his colleagues their chance. It set up an overall organization for oceanography, research centres at Brest, Nantes and Marseilles, and a permanent cadre of oceanographers; it provided for development of Cousteau's 'underwater house', as well as oceanographic ships, submersibles and automatic buoys. Oceanography was designated for concerted action towards 'objectives of national interest'.

The homeland of Lavoisier, Pasteur and the Curies had slipped behind in research compared with, say, Britain, during the 1940s and 1950s, when nuclear energy, antibiotics, computers, rockets, and a torrent of other innovations were summarily ushering technological issues into the cabinet rooms of the world's capitals. Even in the late 1960s research in biology, medicine and agriculture remained underfinanced in France. Yet, despite the post-war instability of governments, French engineers and industrialists had achieved much—most memorably in transport and aviation, hydro-power, and preliminary work on nuclear energy. On the economic base thus created, the ministers and planners of de Gaulle's regime erected a programme matching their wish for global influence.

The programme was far-sighted. In 1964 a study group of the Commissariat du Plan presented the government with a long-term forecast for 1985. The idea was 'to clarify the general orientations of the Fifth Plan'. For example, it envisaged that, despite population growth, consumption per head would be multiplied two and a half times in twenty years, that expenditure on health, transport and leisure would increase much more rapidly than expenditure on food and clothing; that there would be twenty million cars on the roads; that the number of farm workers would be halved; that planners should expect increases in accidents, nervous depression, juvenile delinquency and drug addiction.

The 'committee of twelve wise men', as the French called their government's advisory committee for scientific and technical research, identified work requiring concerted action. Altogether fifteen major projects of this kind were selected for the Fifth Plan—as diverse as automation, town planning and molecular biology. Yet the French concerted actions should not be mistaken for massive crash programmes. Their chief purpose was institutional—to provide support for certain activities in research and technology without the impediments and competitiveness of the pre-existing system for distributing research funds. In short, the concerted actions by-passed the committees of entrenched professors, at a time when the budget of the Centre National de la Recherche Scientifique was increasing at 17 per cent a year.

Like most of the world in the mid-1960s, France suffered from a sense of inferiority towards the United States in the manufacture and use of computers. Two significant events had been the acquisition of Machines Bull, the French computer-makers, by General Electric of the USA, and an embargo by the US government on the export of a powerful computer to the French Atomic Energy Commission, which would have helped the French to make their H-bomb. A special effort was made to improve the situation in computing, again by concerted action and the *plan calcul*. A nationwide network of computing facilities was envisaged, based on three major centres in Paris, Grenoble and Toulouse.

French policy could be excellently collaborative when it suited. France was the first country to commit itself to participation in the proposed giant machine for high-energy physics research in Europe—the '300-GeV' project. The second stage of the Europa launching rocket was made in France. The development with Britain of the Concorde and other aircraft, the agreement of both the Americans and Russians to launch French satellites, a joint project with the Germans for a communications satellite, and participation by French physicists in experiments with the Soviet high-energy machine at Serpukhov—these were just a few of France's techno-diplomatic arrangements.

On the other hand, long discussions in the mid-1960s among broadcasting authorities, aimed at agreeing on a common transmission system for colour television for all Europe, so that programmes would be readily interchangeable, were frustrated in the end by French refusal to adopt a German system (PAL) accepted by the rest of Western Europe. Perhaps significantly for cultural exchanges in the future, the USSR and East Europeans adopted the French system (SECAM).

For a more expressively nationalistic manifestation of French policy, consider the small but significant matter of de Gaulle's instruction to French scientists that they were to insist on speaking French at international conferences. Now, one of the facts of scientific life is that broken English has become almost universally accepted as the international language. The French president did not like it at all, yet his idea could only impair foreign appreciation of French achievements.

Determined French administrators could achieve impressive

results in the application of science and technology and, though the ends they served were the raw ones of national and economic advancement, there was no evidence that anyone wanted it otherwise. On Right and Left there was jealousy of American technological success. The only real arguments were about which leader knew how best to match it.

Even the 1968 students' revolt, despite its element of challenge to the basic goals of the modern technological state, served at one level to accelerate university modernization that would make the technological state more efficient. At a conference in Caen in 1966, representatives of higher education, research and the government had already agreed on sweeping reforms, to break up the centralized structure of the French university: in this respect the students endorsed what the minister of education had already accepted. Jacques Ellul (1964) had been bluntly fatalistic about the monolithic technical world he saw coming into being: 'It is vanity to pretend it can be checked or guided.' But in starting his campaign for the general election forced on him by the student revolt, de Gaulle (1968a), while echoing this fear, shrewdly went on to borrow language from the students. He called for a mutation of French society, through 'participation' that would change the condition of man in the midst of modern civilization. A few weeks later, though, France detonated its first H-bomb; de Gaulle (1968b) hailed the success, 'achieved for France's independence and security by an elite of her children'. His chosen technologies would constrain the condition of man long after his resignation in 1969.

MADE IN JAPAN

For the Japanese, an American toy locomotive brought by Matthew Perry to Yokohama began a developmental experience that culminated in the *Little Boy* A-bomb carried by Superfortress to Hiroshima. By the late 1960s, their country of 100 million souls was firmly established as an outstanding industrial nation that could easily become the third 'super-power' if they chose that role. Lacking the world-embracing zeal of the

Americans or Russians, or the nostalgic habits of industrial enterprise of Europe, the Japanese seemed able to outpace them in economic growth without destroying or abandoning their national customs.

The success of Japan makes it, in some people's judgement, a model of development for the poor countries of the world. So perhaps it does, but it has taken Japan one century to catch up. On that timescale, projected from now into the 21st century, one may doubt whether world conditions will be comparable, or even whether the aims of development will remain the same. Nevertheless, Japan in the 1960s was a fascinating case of a country that had been a technological follower for a hundred years, seeking at last to develop the scientific base that would enable it to be more of an originator. The Japanese having to change was a sign that the character of industrial technology was changing, as its dependence on research became more intimate.

In 1967, work began on a huge 'research park' of Mt Tsukuba, fifty kilometres north of Tokyo; by the late 1970s there were to be 52,000 scientists and engineers working there, in 'several' universities and about forty government laboratories. Contributions to knowledge from Japan continued a tradition in theoretical physics, but also branched out to semiconductor phenomena (the tunnel diode), electron microscopy, biochemistry, microbiology and earthquake studies. Research efforts in industry soared in the 1960s. The Science Council of Japan emerged as the most 'democratic' of bodies for formulating research policies, because the members are elected by working scientists.

The policy of parasitism for countries and companies, that would attempt no original research but would make only whatever scientific effort was necessary for exploiting innovations from outside, had ceased to be effective. The reason was more than a matter of self-respect, although that was a good motive, too. Research had become an essential part of the self-education of an industry, as it was for the self-education of university teachers. While it might not matter much in the end whether an invention originated with yourself or someone else, you had to be well enough informed by your own research to respond quickly but sensibly when an opportunity arose. The important word was 'quickly', because by the 1960s it was apparent that, in the science-based industries, the profitable market life of a particular kind of product could often be shorter than the time taken to develop it.

In retrospect, Japan nevertheless illustrates the separability of three functions in the cultivation of technology: research, innovation and exploitation. During the 1950s, the Japanese were doing far less research than were the Americans or British, but their economic growth was much faster. Nor were they particularly strong in being the first to bring a novel product to the market. Their great strength was in exploitation, using their abundant capital and labour to make the product more profitably. Such was the pattern from shipbuilding to transistor radios. It was to be broken by undersea technology and computers.

MOON OR PLANETS?

What Americans interpreted to be the challenge to their prestige in space was taken up by John Kennedy in earnest, in mounting the huge *Apollo* project for putting men on the Moon. At the same time the Americans, like the Russians, were running military space programmes, and also developing meteorological and communications satellites of direct benefit to earthbound men. But through the 1960s it was *Apollo* that dominated space politics in the United States. By the middle of the decade, technologists and politicians began urgently asking, arguing about, and leaving unsettled, what was to be done with the great resources of rockets and skills being created, after the consummation on the lunar surface. The commitment to *Apollo* remained.

At 'Moonport USA'—Merritt Island by Cape Kennedy, Florida—a crick in the neck told more than any audited expenditure tables, about where the billions of dollars went. Here was thrown up the biggest building of all time, enclosing a greater volume than the Great Pyramid of Cheops, standing on long piles to find bedrock under the swampy land. This temple of the Moon was named prosaically the Vertical Assembly Building and it was designed to house four Saturn V rockets during assembly, with room to spare for rockets even taller than the 364-foot Saturn. Giant precision-built boosters came by barge from the plant at New Orleans, after a visit up the Mississippi to a test-firing site. An assembled rocket trundled upright on a huge crawler for three and a half miles to one of the launching pads—or sacrificial altars. On ignition, the great booster burned for the specified number of seconds, delivering its 3000 tons of thrust; while the upper stages and the spacecraft continued on their way, the spent booster then dropped into the Atlantic Ocean. And that was just the start of the most expensive ride known to man.

Such a ceremony was for the engineers and logistics experts, with the astronauts as the select initiates and research workers mere bystanders. 'Okay, but don't call it science!' was the plea to Kennedy, when he decided to send men to the Moon. But to a public unable to tell the fruits of research from the cultivation of research, the distinction was too subtle.

In any case, the public heard the good news about space in terms that would make a deodorant advertiser blush. 'We have made more discoveries in space medicine that have relieved human misery in the last five years than we made in medicine in the preceding fifty,' said Hubert Humphrey (1966). The vice-president of the United States had been misinformed, to put it mildly.

Serious space research was last in line at NASA. Not that the effort was small, by normal research standards, but when Congress squeezed the budget the Apollo core remained and pips of research dropped out. Caught between Soviet inscrutability and Congressional austerity, the Washington space planners were hard put to it to play the game aright.

Their dilemmas were made plain, just ten years after *Sputnik I*, by the Russian *tour de force* in parachuting *Venus IV* through the atmosphere of the planet Venus. That marked the start of the on-site reconnaissance of the planets, regarded by many as the primary objective of space exploration, especially in view of the immense research interest in the search for life in the unfriendly-looking environments of Mars and Venus. Earlier *Mariner* 'flybys' of both the planets, and Soviet failures, had given the American research workers an advantage. While *Venus IV* was landing, yet another *Mariner* was squinting at Venus from close at hand. It was tantalizing for Americans to know that a probe of theirs could have landed, at that opportunity, had earlier budgets allowed it.

The Jet Propulsion Laboratory at Pasadena had seen its plans for *Voyager*, a big, versatile probe to explore the solar system and make some landings, delayed and delayed again. Leading research workers recruited to the space effort were dispirited when the flight of the first *Voyager*—intended to land on Mars—was postponed until 1973. Before then, their Soviet colleagues would be quite likely to make any scoops allowed by nature, concerning extra-terrestrial life and the surfaces of the near planets, And, by then, too, the Moon might well seem pretty dull by comparison.

James Van Allen (1967) of the University of Iowa, who made the first major space discovery—that of the radiation belts around the Earth—demanded to know, 'Are we to abandon the planets to the Soviet Union?' The Russian bogey was still doing yeoman service in the defence of science.

DAYS OF RECKONING

Back in the Kremlin—or, to be more precise, across Manejnaya Square in Gorky Street—the deputy prime minister responsible for science and technology had little to be complacent about. To be sure, Vladimir Kirillin could claim, as he said to me in that office, 'Successes in space research are in a sense a certificate of maturity for our Soviet science'. So, in a sense, they were. You could not fly men in orbit or probes to Venus without a great deal of competence in aerospace engineering, metallurgy, electronics, and so on. But competence and excellence, especially in research and technology, are really very different things. As an academician and Lenin-prizewinning physicist, Vladimir Alexeyevich would know that as well as anyone.

The dearth of excellence in the Soviet scientific landscape made American agitation about Russian prowess ironical. Even by the late 1960s, and excluding only the space effort, the discoveries and inventions coming out of civilian laboratories in the USSR did not measure up to the current achievements in the UK, never mind the USA. Officially, no government had more consistently or for longer dedicated itself to the cultivation of research and technology than had the Communist regime in Moscow. An army of research workers and engineers, matching the Americans' in numbers, had been created in a huge pedagogic programme. Yet it was an army without élan.

In some of the higher reaches of physics and of astrophysics, the Russian long-standing love of mathematics paid off frequently. In the important discoveries of the maser and the laser, Soviet laboratories were well to the fore, and the Swedish Academy of Sciences did justice in dividing the Nobel prize for the maser between the American Charles Townes on the one hand, and Alexander Prokhorov and Nikolai Basov on the other. Other fields where the effort looked impressive included exploration of the solid earth and of the oceans, and surgery. Where large scientific projects were required—notably in building the 70-GeV high-energy machine at Serpukhov, or the world's biggest optical telescope in the Caucasus—the Moscow planners would oblige. And sometimes the Soviet researchers would take an opportunity that their Western colleagues missed. For example, they adapted a laboratory technique using fluorescent antibodies to the rapid clinical identification of viruses.

Such bright spots merely contrasted more sharply with the drabness of the general scene. In big areas like chemistry and biology, friendly Western visitors repeatedly returned home with thumbs turned down. At the new growing points—radio astronomy and molecular biology, for example—contributions from the Soviet Union caused little stir. Kapitza (1966a) described the situation as mildly as possible when he wrote: 'We must not be afraid to say that, in the last few years, the gap in science between this country and America has not been closing.'

That would have been enough for Kirillin, deputy prime minister, to brood about in his big office on Gorky Street. But there was much more than a cultural failure to account for. By his party's doctrine science was a productive force. Yet the American capitalists showed themselves incomparably better able to seize the new idea from research or technology and apply it quickly in production than did the managers of Soviet industry. Moreover, although Soviet research workers were ceaselessly urged to be useful, it was not they who had conceived the jet engine, or the new electronics of the solid state, or the post-war style of petrochemical industry, or the vital new drugs. Time after time these inventions had to be imported, belatedly, from the West and little went the other way. There were again bright spots, to be sure, in metallurgy, machine tools and power generation, as well as military and aerospace technology. But agriculture and most other industries were embarrassingly backward.

A group of Western experts analyzing Soviet science policy (OECD, 1969) noted frequent organizational upheavals but persistent isolation of leading research institutes from both students and factories. A French member of the group, Eugène Zaleski, wrote of the early 1960s: 'the Soviet Plans for Science and New Technology on a national level were very badly executed.'

Most vexing of all was the lack of advanced electronic computers. I once tried to persuade Kirillin to say something to a BBC camera about 'cybernetics'. He evaded the word persistently, and so provided a mute commentary both on politics and on computer manufacturing in his country.

'Cybernetics' had figured prominently in the new Communist Party programme introduced by Khrushchev in October 1961. It was the term used by the American mathematician Norbert Wiener in generalizing and popularizing post-war ideas on automatic control and the uses of computers. With the passing of Stalinism and the launching of the first *Sputnik*, leading Soviet research workers and technologists had found in the word a new ideological weapon to use against political dogmatists who were cramping their style. In the late 1950s the new 'super-science' of cybernetics was hailed by academicians and political reformers, with backing from the party praesidium, as the means of modernizing industry and reinforcing scientific development. 'In future, a new type of national economy will arise in which automated production will predominate,' declared the official textbook, *Fundamentals of Marxism-Leninism,* published in 1959. The leading radio engineer Axel Berg announced that ' . . . under socialism it is perfectly possible to organize a single complex automated system of control of the country's national economy' (Paloczi-Horvath, 1964).

By the mid-1960s the fervour for cybernetics was spent. Soviet experts had become as chary of using the term as their colleagues in the West; it smacked of excessive optimism, if not of slight crankiness. Worst of all, the creed of cybernetics seemed like a bad joke when many of the nation's computers were museum pieces and even the big new Ural-4 was obsolete at completion. Victor Glushkov (1966), the able head of the Institute of Cybernetics in Kiev, wrote frankly of difficulties with computers, and of Berg's vision remarked: 'The impression that we can now control the national economy by computer unfortunately is not correct.'

BLIGHTED BIOLOGY

There were other difficulties for the Soviet science planners, of a more starkly political kind. If the pre-war judgements of official philosophers had been heeded, and Soviet physicists had dismissed as bourgeois the theories of relativity and quantum mechanics, Igor Kurchatov and his colleagues could not have developed nuclear weapons. As it was, Joseph Stalin's endorsement of Lysenko's doctrine of the heritability of acquired characteristics—well matched to ideas of human perfectibility—had driven Soviet biology into fantasy for a generation. It damaged Soviet agriculture at least as much as did collectivization or Khrushchev's dust bowl in the Virgin Lands.

It was no use saying that Trofim Denisovich Lysenko was a rogue. The system had backed him, desperately wanting his magical solutions to stubborn agronomic problems. At a time when Soviet farmers needed the benefits of scientific plant and animal breeding, they were treated to false claims, phoney products and escalating promises. If Lysenko had a trick, it was to break the rules of research so barefacedly and frequently that conscientious scientists had no time to refute him in their own terms. Proper research on plants and animals is a slow business, tied to the seasons and requiring years to secure reliable results; before one of Lysenko's announcements could be checked there would be half a dozen more from the same source.

After a while he did not need to bother about tricks. He and Stalin were thorough in their planning of Soviet biology. Geneticists adhering to the ideas of Gregor Mendel were imprisoned, sent to the front, sacked, or switched to other work. A pitiful Mendelian underground persisted under different subject names. All teaching and research in genetics not under direct Lysenkoist control was officially stopped. The best geneticist in post-revolutionary Russia, Nikolai Vavilov, died in a prison camp.

At an ideological conference at Brno, Mendel's home town, Czechoslovak scientists conformed by publicly repudiating Mendelism. That was in 1951; fourteen years later, in the same place (with the same academician in charge), they restored it. In the meantime, Mendelian genetics had continued fairly well, in a concealed fashion. As a Czech scientist remarked, 'Lysenkoism was a catastrophe, but not very serious!'

It was serious enough in the USSR. In February 1965 the Academy of Sciences finally and irreversibly rid itself of Lysenko. It was a good gesture then to create a prize in honour of Vavilov, but there were few to compete for it, except survivors from the early 1930s and the brave spirits who had contrived to carry on with genetical research under the guise of biochemistry, radiation biology, agriculture or medicine. A generation of research workers was almost missing; there was not a text-book in sight.

The new director of the Academy's Institute of Genetics,

Nikolai Petrovich Dubinin helped to pioneer modern population genetics before taking refuge in ornithology at the height of his predecessor's power. In his late fifties he took on the task of filling the vacuum. He was to build up twenty laboratories, including an Institute of Cosmic Genetics. The Moscow Natural History Society published a special Mendel Memorial issue of its proceedings. At international meetings the Soviet participation confirmed that the greatest barrier to East-West scientific understanding had fallen. The new journal *Genetika* carried learned papers in the international idiom. The recantation was complete and with Mendelian biology Soviet researchers began to secure results of the kind that the Lysenkoists had obtained only by wishful thinking.

So much for biologists. But no researcher works in an intellectual vacuum. A young Moscow physicist, Pavel Litvinov, grandson of the former foreign minister, was sacked from the Institute of Precision Chemical Technology because he publicly protested about a trial of underground editors (he was later sentenced for further offences). In an underground pamphlet (1968), Andrei Sakharov, one of the younger academicians, denounced the science chief at the Communist Party's central committee, Sergei Trapeznikov, as a key neo-Stalinist. As long as Trapeznikov continued in power, Sakharov wrote, 'it is impossible to hope for a strengthening of the party's position among scientific and artistic intellectuals'.

The alternative? Well, the Czechs had not stopped at rehabilitating Mendel. Out of the Academy of Sciences in Prague had come fiercely revisionist economic notions and new ideas about freedom. They recognized the correlation between science and democracy, and the 'subjectivity' of creative science. Eventually they toppled the President, Antonin Novotny, and the Czechoslovak Communist Party (1968), thick with scientists in its leadership, went so far as to declare modestly that it was not the proper instrument for the dictatorship of the proletariat. Such was the heresy yawning before the Soviet leaders in the scientific age, to which tanks were an inadequate answer. It was cosier in Lysenko's time.

THE LEARNED PATRIOTS

When Albert Einstein (1950) complained that 'The man of science has slipped so much that he accepts the slavery inflicted upon him by national states as his inevitable fate', slavery was not quite the right word. Scientific leaders 'sold' research to rulers by appeal to national advantage. They sought the bribe.

Since Daedalus worked for Minos, discoverers and inventors have needed patronage and have therefore bent their interests to those of competitive men. If they failed to do so, then or now, they could not work effectively; they also failed to serve their communities, by the conventional tests of social duty. Einstein himself was one of those who urged Franklin Roosevelt to develop the A-bomb, because of the danger that the Nazis would get it first.

Three decades later, the moral status of national purposes was much less plain. When a biophysicist, Ernest Pollard (1967), appealed to his fellow American academics to bend their minds to ways of winning the war in Vietnam, he had a chilly reception. After two months, some 180 scientists had enlisted in his 'VN Emergency Science Group' but the general run of opinion in the university laboratories was by no means happy with US policies in Vietnam, let alone wishing to contribute research services in support of them.

By then it seemed much too late for such scruples. Warlike science had become a habit. John Kennedy's administration, abetted by the very experts who shrank from the suicidal folly of nuclear war, began developing the new capacity for limited and counter-insurgency war that was to be first put to the test in Vietnam.

Provision of a supply of bright young men, to staff the nuclear weapons establishments, the military electronics laboratories, the missile design offices, the chemical and biological warfare stations—this was part of the moral tax that science paid for its prosperity. It seemed that the rich, in defence of national advantage, would forever shower napalm and herbicides and toxic gases (and their future successors) on dissidents in the underdeveloped world, when what they needed was protein and energy and life-saving drugs. Here was the natural consequence of the work of the learned patriots, that the dominant promise of the scientific age had become death by violence.

The implication of the scientific community was symbolized by the presence in the US Department of Defense at the height of the Vietnam War of Frederick Seitz, president of the National Academy of Sciences, doubling as chairman of the Pentagon's defence science board. Seitz's successor as president of the Academy in mid-1969, the biochemist Philip Handler, indicated no readiness to break the ties between the Academy and the Department of Defense, though he was plainly anxious that the Academy should take more initiative about domestic problems.

A British parallel seems at first sight innocuous: that for several years the president of the Royal Society, Patrick Blackett, was simultaneously special adviser to the minister of technology. In London in the late 1960s, the application of research and technology to bolstering the national economy, and in particular to boosting exports, appeared as natural and compelling as the quest for centimetre radar had done in the 1940s. There might be misgivings about whether emphasis on money-making applications of discoveries and inventions would reduce support for long-term research, because voices were heard saying that there was 'too much pure science in Britain'. But few questioned the central economic motives or industrial ambitions, or mocked the absurd situation developing amongst the rich nations.

All technologically advanced countries were dedicated to a mutually self-defeating policy of exporting as much as they could in the way of sophisticated products, and importing as little as possible. Civilian technology was sponsored primarily according to its marketability, not its contribution to a better life. Civil research was becoming as subject to the salesman as military research already was to the general. More profoundly, technology had given a new reason for the existence of nation states. As Jean-Jacques Salomon (1966) put it, the new society was concerned 'no longer only to produce what has been found, but to find what one wishes to produce'. There is a circle: nations exist to create the technology that nations need to exist.

The technological competition between nations conceals or overrides the opportunities for technological choice within a nation, and for social control of the uses of science. In other words, there is a market-place consensus between the nations, by which technologies are promoted primarily on grounds of possible overseas sales, preferably to hard-currency countries.

Although the tradition of research is one of international collaboration, the tradition of technology is of competition. Over a very wide range of technologies, from textiles and steel-making to aero engines and electronics, the competitive spirit grows stronger. One can check off the developments that threaten to segregate the world in hotly competing nations or alliances rather than bring them together: they include nuclear power technology, aerospace technology, ocean technology, big computers and, of course, weapons. Add to those the emphasis in chemical technology on devising synthetic substitutes for imported products, and the conclusion must be that the most substantial uses of science are at present dividing the world into competing arsenals and supermarkets wearing familiar national flags.

CHAPTER 3

Men of the World

ABDUS SALAM likes to recall tales from the time when Islam was the think-tank of the world and the brain drain flowed to Baghdad or Moorish Toledo. He finds particularly poignant the story of the aged philosopher and historian Ibn Khaldun who in 1401, regardless of personal safety, had himself lowered in a basket from the walls of beleaguered Damascus, to plead with the Tartar emperor, Timurlane, that the city should be spared. To everyone's amazement, Ibn Khaldun's mission succeeded and the butcher of Baghdad, Delhi and Moscow left Damascus in peace. Salam, whose own name means 'servant of peace', wishes that the natural philosophers might have the same courage and skill to turn aside the wrath of national leaders in the thermonuclear age.

For Salam, the Punjabi country-boy who became a 'tribal chief' in the most mind-wrenching branch of learning, it is entirely natural to travel frequently by intercontinental jet, to keep up with his commitments in research and public service. He personifies the international spirit of scientific research and that world-wide network of interchange and co-operation between research workers, regardless of nationality, which contrasts sharply, almost paradoxically, with the application of so much of modern knowledge to military and economic contests. Salam's home is in London, where he is professor of theoretical physics, but you are as likely to find him on the Adriatic coast at Trieste, directing the International Centre for Theoretical Physics; or in Rawalpindi, where he was scientific adviser to President Ayub Khan; or at the United Nations, as a member of the advisory committee on the application of science and technology.

The Trieste centre was Salam's idea and it was set up with

support from the International Atomic Energy Agency and the Italian government in 1964. Theorists from Asia, Africa and Latin America come to Trieste to make contact with the ideas current in Europe and North America, and lose their sense of isolation by conversing with world authorities. They thereby escape the fate of Salam himself as a young professor in Lahore, who had to choose between stagnating in an intellectual backwater or deserting his homeland. Physicists from Eastern Europe and the USSR can work together with physicists from Western Europe and the USA—as they did, for example, in 1965 in charting, in a year-long study under the joint leadership of an American and a Soviet physicist, opportunities for research towards peaceful thermonuclear power. Salam calls the centre 'the first department of a UN university'. In June 1968 he brought three hundred leading physicists from many countries to Trieste, for a stocktaking of the whole subject.

The physicists have no monopoly in interchange. While that meeting was in progress in Trieste, the botanists, for example, met at Freudenstadt in Germany to discuss photosynthesis; the eye surgeons assembled in Amsterdam; world experts in chemical catalysis were in Moscow; molecular biologists debated the structure of viruses in Cambridge, England. Nor is it just in such meetings or in the frequent travels of research workers that the strength of everyday international relations is manifest; look in any scientific library and the range of foreign books and journals testifies to a continual transfer of ideas and information. A typical issue of the journal *Nature*, published in London, contains research reports from Dunedin, Genoa, Chandigarh, Osaka, Oslo, Leningrad, Basel and Namur, as well as from various British, American, Australian and Israeli centres. A French expert on whales may never have learned the name of the prime minister in Tokyo, but he will know very well who are the top Japanese authorities on the cetacea.

Love is scarcely too strong a word for the understanding that persists among research colleagues whatever the political divisions between their nations. Laws of physics are invariant even where social laws differ widely; medical discoveries will save life under any flag; a new technology can generate wealth indifferently for socialized industry or for private profit. From the very origins of modern science, overseas correspondence was regarded as essential for the pooling of knowledge. The need to do certain researches in particular geographical locations provides another motive for co-operation; so does the wish to share the costs of experiments. Even the brain drain is proof of the international character of scientific knowledge. No other profession enjoys the ready intercommunication of scientific research.

There are difficulties. With some countries, foreign-exchange shortages or political circumstances inhibit communications. The Moscow bureaucracy frequently confounds conference organizers who invite Soviet participants. The suitability of NATO countries for scientific meetings was called in question when East Germans were refused travel permits. Yet, at a time when all other communication with Peking had virtually ceased, during the Cultural Revolution, the Royal Society of London continued for some months sending visitors to the Academia Sinica. And even when the illegal regime in Rhodesia brought ostracism to that country, scientific contacts persisted quietly but fruitfully in research programmes relevant to the welfare of Africa as a whole.

On rising ground, straddling the frontier between France and Switzerland near Geneva, stands the masterpiece of international scientific enterprise. At the laboratory of European Organization for Nuclear Research (CERN) high-energy physicists of many nations have worked together amicably and fruitfully since the 1950s, shredding nuclear matter into fine pieces to try to discover the laws that govern it. Thirteen Western European governments paid for a big (28-GeV) high-energy machine for the purpose, and have since provided computers and advanced instruments of many kinds—enabling their physicists and those of other countries to carry out research that would be thought too expensive for any one European country to sustain on its own. CERN averted the risk of the eclipse of Europe in one of the most exciting areas of fundamental science. For anyone with a sense of twentieth-century history, to see Frenchmen, Germans, Italians, Englishmen, Dutchmen, working together forgetful of nationality, has a value additional to the quality of the work itself. When the European physicists demanded a new accelerator ten times more powerful and expensive, to be built under a similar arrangement (the 300-GeV machine), the finance ministers mopped their brows and raised objections to the scheme; but no one could question the physicists' ability to make the collaboration work.

CERN was founded on a misapprehension. In the early years after Hiroshima, the world's statesmen were understandably impressed by the new power put into the hands of nations by the physicists. Two very different interests coincided; that of the physicists, who were anxious to exploit a wartime discovery of a new principle for big, but expensive, accelerators; that of politicians anxious to find some suitable exercise to display European unity. As Lew Kowarski (1961) put it: 'Atomic energy was attracting public readiness to spend money.' Uranium fission itself was taboo for international co-operation at that time; the physicists succeeded in getting the best of both worlds by using the 'nuclear research' tag while explaining that high-energy physics was not a secret subject.

The political assumption was that further abstruse investigations of sub-atomic matter would yield new practical benefits, and that Europe could not afford to fall behind the USA in such research. In the preamble to the CERN agreement of 1953, the purpose of the international laboratory was: '. . . to study phenomena involving high-energy particles in order to increase the knowledge of such phenomena and thereby to contribute to progress and to the improvement to the living conditions of mankind'.

There were, and are, fields of nuclear research that definitely hold out promise of 'improvement to the living conditions of mankind'—notably, research towards peaceful thermonuclear power—but the kind of physics done at CERN is probably not in this category. If none of the participating governments was hoodwinked in that preamble, they were at least misleading their own taxpayers. No discovery with the big machine at Geneva contributes, or is ever likely to do so, one centime, lira, pfennig or penny to the material well-being of Europe. The CERN laboratory is dedicated to the enlargement of knowledge for its own sake and to the discovery of pheomena that are in a quite real sense 'out of this world'.

The most publicized world-wide adventure in research, the International Geophysical Year 1957–8, coincided with a period of extreme storminess on the Sun. The Russians announced their first *Sputniks* as 'contributions' to the IGY. But with many nations' laboratories, stations and expeditions committed to simultaneous programmes of research on the Earth, its atmosphere, its oceans and its polar wastes, the IGY had already caught the imagination of physicists and public alike. The main reason for the co-operation was that various natural phenomena, particularly those connected with storms on the Sun, could be understood only with a global picture of what was going on. This pattern of collaboration had begun with the International Polar Years of 1882–3 and 1932–3. The pace quickened and the IGY was followed first by the International Years of the Quiet Sun (1964–5) and then, as the results of that programme were being digested, a new, continuous operation was instituted under an inter-union commission of solar-terrestrial physics. Like the bodies that master-minded the IGY and the IQSY, it is an arm of the International Council of Scientific Unions—ICSU, the 'scientists' UN'.

The intensive scientific activity in Antarctica during the IGY paved the way for political immunization of the continent. Interested nations agreed neither to renounce nor press any territorial claims and left the continent open to exploration under the Antarctic Treaty of 1959. Some nations, as a result, relaxed their research effort, as they no longer felt obliged to show the flag, but others did more, notably the Japanese. Another ICSU body, the Special Committee on Antarctic Research (SCAR) provides the framework for ongoing co-operation in research. The chief foreseeable significance of Antarctica in practical affairs will come when airliners take the south polar route in intercontinental journeys, as they already take the north polar route; then a refuelling field, emergency landing strips, radio navigation aids and more permanent weather stations will appear on the continent and its surrounding islands. Fortunately, although Antarctica has the world's largest coal-fields, all geological exploration has so far failed to find economically accessible deposits of minerals. The treaty negotiators were unable to decide how to deal with any such development and it could be disastrous for the agreement.

In a sense, international co-operation in research was too successful. Organizations and agreements proliferated, until national science policymakers began to fear that they were losing control of their own programmes, by long-term commitment to international enterprises. Nor were all the collaborations as successful as CERN and the IGY. Some grand schemes were stillborn and others faltered for lack of support by the nations. Nevertheless, the internationalism of research remains one of the most hopeful factors both for world peace and for world-wide control of the uses of science.

THE PEACEMAKERS

Big developments need big men, in weaponry as in any other field; they usually turn out to be cultivated, warm-hearted people. Those who made radar and the A-bomb during the Second World War were the cream of a scientific generation of Europe and North America, many of whom showed passionate concern for the welfare of mankind. With the H-bomb were associated such cheerful men as Stanislaw Ulam in the USA, William Penney in Britain and Igor Kurchatov in the USSR; with napalm, Louis Fieser, professor of chemistry, who first tried out his sticky gasoline on a Harvard sports field; with the Polaris submarines, Hyman Rickover, admiral extraordinary. Some of the former weapon-makers, notably Edward Teller and Wernher von Braün, people try to cast as Dr Strangelove, hell-bent on destruction or utterly indifferent to humanity; yet it is hard to make them fit the part, however much one may quarrel with their outlook or judgement. Indeed, anyone hunting for the mad scientist of the caricatures not only will fail to find him but will fail to assign correctly the responsibility for current military trends and the race for new weapons.

> 'Scientists have often before been accused of providing new weapons for the mutual destruction of nations, instead of improving their well-being. However, in the past, scientists could disclaim responsibility for the use to which mankind had put their disinterested discoveries. We feel compelled to take a more active stand now because the success we have achieved in the development of nuclear power is fraught with infinitely greater dangers than were all the inventions of the past.'

One of the remarkable documents of science politics is this report by researchers in Chicago engaged in the A-bomb project (Franck, 1945). It is dated nearly two months before the nuclear attacks on Japan. The chairman of the group was James Franck, a German refugee who was a leader of chemical work on the A-bomb. Other members included Leo Szilard, Hungarian-born physicist who had been one of the initiators of the project, Glenn Seaborg, American-born discoverer of plutonium who, two decades later, was to be chairman of the US Atomic Energy Commission, and Eugene Rabinowitch, Russian-born, German-educated biophysicist and subsequent editor of the *Bulletin of the Atomic Scientists*.

In their preamble they explained the need to discuss the global political implications of the weapon under development. Franck and his colleagues went on to dismiss the idea that the United States could avoid a nuclear arms race by keeping scientific facts secret or by cornering the uranium ore. They greatly underestimated the power of the A-bomb—the Trinity test explosion did not take place until the following month—and also the likely increase in the power of nuclear explosives during the decade ahead. Even so, they argued that it would be difficult to make the USA safe from nuclear attack.

'If no efficient international agreement is achieved, the race for nuclear armaments will be on in earnest not later than the morning after our first demonstration of the existence of nuclear weapons.' They estimated that it might take another nation three or four years to obtain the bomb. That was not a bad guess because the first Soviet test was in 1949, though in fact the Russians had already begun work in 1943. The Franck committee also anticipated that an international agreement on prevention of nuclear armaments would have to be backed by efficient controls, hinging on book-keeping for the conversion of uranium ore into a high-grade fissionable material. That was just the kind of safeguards system still envisaged in 1969.

To prevent mistrust springing up which would impede international agreement for avoiding nuclear warfare, the Franck report urged that the new weapon should not be used directly against Japan, but should be publicly demonstrated in the desert or on a barren island. It wanted the other members of the United Nations to share the responsibility for the decision about whether the bomb should be subsequently used against Japan. The War Department asked four leading physicists—Arthur Compton, Enrico Fermi, Ernest Lawrence and Robert Oppenheimer—to devise a non-lethal demonstration that would infallibly frighten the Japanese warlords into surrender; regretfully, they reported: 'We can see no acceptable alternative to direct military use.'

The stories of how, as the war ended and in the early post-war years, the 'atomic scientists' strove for rational political responses to the existence of nuclear weapons have often been told before. The Franck report was just one effort among many. Winston Churchill had lost patience when Niels Bohr of Copenhagen, the most eminent physicist in the world, went to plead with him in May 1943 for international control of nuclear energy; as a last resort, Bohr asked if he might send a memorandum and Churchill replied that he would always be honoured to receive a letter from Professor Bohr, but hoped that it would not be about politics. (Gowing, 1964). Despite such rebuffs, the scientific community, from 'establishment' figures like Vannevar Bush and James Conant of the USA and Henry Dale and John Cockcroft of Britain, through to the more rebellious spirits like Szilard and Patrick Blackett, took the political consequences of nuclear energy immensely seriously and the almost unanimous call was for disarmament under international inspection and control.

The McMahon Act of 1946 was a triumph for the American scientists; it put nuclear development under the civilian control of an Atomic Energy Commission. Even Truman's much resented decision to withhold nuclear information from London, disregarding agreements to the contrary and the substantial British initiative in the development of the A-bomb, was seen by Americans as a first step in arresting the spread of nuclear weapons. In London it seemed more of a provocation. Clement Attlee (1967) said of his decision to make the Bomb: 'We'd just got to show them they didn't know everything.'

The conclusion of the partial test-ban treaty of 1963 was due in large measure to the efforts of research workers: to the agitation about the hazards of radioactive fall-out, from Linus Pauling and others in the United States and from Joseph Rotblat and others in Britain; to the scientific experts of East and West who met in Geneva to agree on the technical feasibility of detecting nuclear tests; to the scientific advisers in Washington, London and Moscow who threw their influence in favour of the treaty in the face of opposition from the military.

Cockcroft, first begetter of both military and civil nuclear energy programmes in Britain, devoted the last days of his life, in the autumn of 1967, to countering the objections from Bonn and Brasilia to the proposed non-proliferation treaty to halt the spread of nuclear weapons. At the same time he was demanding that the existing nuclear powers should cut back their own nuclear weapons programmes.

As Americans had first conceived it, in Eisenhower's Atoms for Peace programme, an International Atomic Energy Agency would provide a world 'bank' of nuclear fuel and technological secrets, to be made available to nations in return for safeguards against military use. The simple idea scarcely survived when the agency came into existence in 1957. There was, after all, no great shortage of nuclear fuel; the 1955 conference in Geneva made many of the so-called 'secrets' freely available; and a second conference, in 1958, showed that nuclear energy was far from economic at that time. The Russians in the early days of the IAEA were inclined ferociously to regard almost every activity of the agency as antagonistic to themselves; the French and the Indians were resentful of the agency's wish to regulate the activities of nations not already possessing nuclear weapons. Yet the agency provided the only serious basis for preventing the misapplication of reactors and fuel to military purposes, and joint US-Soviet proposals for the non-proliferation treaty (1967–8) gave the IAEA the task of inspecting reactors and fuel separation plants. Sigvard Eklund (1968), the IAEA's Swedish director general, warned that by 1980 power reactors in countries not already possessing nuclear weapons would be capable of producing enough plutonium for a hundred A-bombs a week. By 1968, the IAEA inspection system was operating on an *ad hoc* and voluntary basis for seventy nuclear reactors in twenty-nine countries. A treaty declaring Latin America a nuclear-free zone had charged the IAEA with the responsibility for inspection.

Eventually the UN General Assembly adopted the non-proliferation treaty, despite reservations by some member states. Eklund and his staff began to prepare for the first world-wide policing operations in support of a major arms control agreement. It was a triumph for the scientists of East and West who had sustained pressure for such a treaty for many years, even when it seemed a bizarre proposal in a war-torn world.

But the task of hauling the world back from nuclear catastrophe was by no means complete. The goal of reducing the US and Soviet stockpiles of nuclear weapons seemed hopeless, unless, for a start, the new arms race in nuclear-tipped anti-ballistic missiles could be averted. In 1967 Donald Hornig, as President Johnson's science adviser, had firmly opposed Robert McNamara's anti-ballistic missile (ABM) programme, advertised as a screen against possible Chinese nuclear attack and planned for deployment in the suburbs of major cities. Three former presidential science advisers joined in open protest against the plan. One of Richard Nixon's first acts as president in 1969 was to bow to mounting criticism by reducing, rather than expanding, the programme and by shifting the launching sites away from the cities.

Anxious glances then fell on the gas centrifuge, for concentrating uranium-235. Years of secret work ended in an open agreement between the Netherlands, Britain and West Germany, in 1969, to build two centrifuge plants. The intended product was enriched fuel for nuclear power stations. But uranium-235 also serves in H-bombs and the fear was that this technology would speed the spread of thermonuclear weapons to more nations. That so marginal an advance was thought politically risky showed how slender was the thread of human confidence.

THE PUGWASHITES

The Pugwash movement became the principal vehicle for the man of science as peacemaker. In July 1955 the Russell-Einstein Manifesto was released at a news conference at Caxton Hall in London. There were other signatories but, as the manifesto was drafted by Bertrand Russell and signed by Albert Einstein two days before his death, their names provided the label by which it came to be known. It outlined the risks of thermonuclear war, remarking, 'We have found that the men who know most are the most gloomy'. It concluded:

> 'We appeal, as human beings, to human beings; remember your humanity and forget the rest. If you do so, the way lies open to a new Paradise; if you cannot, there lies before you the risk of universal death.'

The manifesto received world-wide publicity, and Cyrus Eaton, a Cleveland industrialist, wrote to Russell offering to finance an international meeting of scientists at his birthplace, the fishing village of Pugwash, Nova Scotia. After an abortive attempt to stage a meeting in India, the first conference duly took place at Pugwash in July 1957. Twenty-two participants, mostly physicists, attended; they came from ten countries, including the USA, the USSR and China. Despite all the expected political difficulties, they reached general agreement on many aspects of world affairs and, where they failed to agree, they were able to define their differences.

During the first ten years thereafter there were sixteen conferences in various countries. All the issues discussed in this chapter were aired in the Pugwash Conferences on Science and World Affairs. In the early period, when even to talk to Russians and Chinese was still regarded in the West as approaching treason, Pugwash had to contend with suspicions of pro-communist tendencies; gradually this illusion was dispelled as men of high standing with Western governments attended and found the conferences useful—men like Jerome Wiesner, George Kistiakowsky and Alvin Weinberg of the United States, John Cockcroft, William Penney and Solly Zuckerman of Britain, François Perrin and Pierre Piganiol of France.

Research workers are by nature inventive, and one of the features of the Pugwash meetings has been the airing of technical schemes. Some, like the 'black box' (sealed recording seismograph) for monitoring underground nuclear explosions, came from official studies. Others, like the UN sonar network, were private ideas of the participants. The latter proposal would limit the mischief made possible by the untagged missile-carrying submarine which, aided by the concealment and free movement offered by the oceans, might provoke war between other nations than its own. As proposed by Japanese and American Pugwash scientists, a world-wide detection system would keep itself informed about what block of ocean each submarine was occupying. Other Pugwashites shook their heads over the idea of the UN becoming engaged in what could be, in effect, a technological arms race with the permanent members of the Security Council.

The long-lasting conversation between East and West at Pugwash did much—how much may never be measurable—to prepare the way for the treaties on the test-ban, outer space and non-proliferation. The participants attended as individuals, but many of them reported back to their governments. People from Peking attended up until 1960; thereafter it was a major concern to try to renew the links with China.

Alert to the expectation of new weapons, Pugwash set up a very practical study group on biological warfare. It not only assessed the risks and the international legal situation but also conducted practical inspections of microbiological laboratories in several countries (first in Denmark, Sweden, Austria and Czechoslovakia) to test possible procedures for detecting illegal manufacture of biological weapons, that could be employed under a disarmament treaty. At the same time international studies began, of means for detecting biological attack (either for defence or in investigation of alleged attacks) with special emphasis on a 'fluorescent antibody' method for identifying disease agents. Chemical weapons also came under review.

Lev Artsimovich (1967) proposed a treaty banning any research and investigation for the purpose of creating new weapons of mass destruction. 'An international scientific-technical committee consisting of eminent worthy scientists, each of whom could be responsible for carrying out the treaty in his field in his own country, could become a guarantor of the fulfilment of such a treaty.' Here was a typical Pugwash idea —at first sight extraordinarily naïve or suspicious. How could a world so full of mistrust rely on the word of, say, a German expert, that no research on biological weapons was proceeding in his country? Why should scientists come to occupy a special diplomatic position, as Artsimovich proposed? Wasn't this just a way of dodging the inspection and control issue, as Russian negotiators had done for years? Of course it was absurd, by conventional diplomatic standards. But Artsimovich, one of the most eminent physicists of the USSR, was no fool, and a moment's reflection indicated that there was a whole range of possible but so far non-existent military inventions that might be prevented from emerging only by a scheme something like his. By the 1980s it would not seem so odd an idea.

The tenth anniversary meeting of Pugwash, at Ronneby in Sweden in September 1967, demonstrated the magic of Pugwash in action: three months after the Israeli-Arab war, Ronneby was probably the only place in the world where Jews and Arabs could be seen eating together, discussing politics. East Germans and West Germans talked of the future of Europe; scientists from the rich countries and poor countries argued about the roles of aid and of social revolution in world development. But it was plain that this 'conversational mode' of Pugwash conferences had passed its point of maximum usefulness; thereafter, working parties were to carry out more thorough and detailed studies of particular problems than in the past.

Joseph Rotblat, a Polish-born medical physicist of London University, had been one of the earliest participants in the wartime A-bomb project, starting in Liverpool in 1939. He was a signatory of the Einstein-Russell manifesto. On him fell most of the responsibility for organizing Pugwash and holding it together through political and administrative crises, during its first decade. He had a full-time university job while he laboured to exhaustion in his spare time as secretary-general of the movement. Pugwash became much the most effective unofficial agency of peace and understanding in the post-war world, and science and humanity are more in Rotblat's debt than they may ever realize.

For social scientists and, to a lesser extent, for natural scientists, there was another, more professional course to a safer world: the serious prosecution of 'peace research'. That meant academic study of the causes and character of conflict, means of resolving conflict and achieving disarmament and the nature and problems of a peaceful world. Most of what might be labelled 'peace research', and by 1969 there was not much of it, was being conducted by individuals and small groups. An Arms Control and Disarmament Agency was set up in 1961 as an official US body responsible for such research, though carefully aligned with government policy and housed in the State Department.

One of the larger non-governmental enterprises is in Stockholm, with a staff of some twenty professionals. To mark 150 years of neutrality, the Swedish government decided to endow an international institute and the Stockholm International Peace and Conflict Research Institute (SIPRI) came into existence in 1966, following consultation with Pugwash. A study of biological warfare, in close collaboration with Pugwash, was first on the SIPRI programme, leaving chemical warfare for subsequent inquiry. Attention to disarmament began with the goals (1) of a comprehensive test-ban treaty, covering underground tests, which meant examining the state of play in seismic detection of clandestine tests; (2) of the non-proliferation treaty, in which connection SIPRI was particularly concerned with the interconnections with the peaceful uses of nuclear energy; and (3) of general disarmament, which seemed to have slipped badly in the diplomatic agenda. Another topic for investigation was the way the press and other mass media reported conflicts and the efforts to secure disarmament. Communications satellites and their use by the super-powers were to be examined as a possible source of international tension. And SIPRI prepared an annual review of the arms race, of the international traffic in arms, and of progress in arms control. A study of ocean law was undertaken.

THE TORMENTED

There seemed to be as much of curse as of prophecy in the words of a British minister for science, Quintin Hogg: 'I remain obstinately of the opinion that, in the long run, the marriage between science and defence is corrupting, and will at best turn science from a liberating to a destructive force, and at worst ultimately dry up the wells of inventiveness in the scientist himself.' (Hailsham 1963)

What are we to make of the contrasting images of the scientist in international affairs: the globe-trotting researcher collaborating as a matter of course with his colleagues of any nationality; the military research worker; the science adviser counselling the defence minister on the development of new weapons; the Pugwashite declaring himself with great sincerity for peace and disarmament above all else? Perplexity may be heightened when the same man stars in two or three roles simultaneously, slipping from one to the other without even stopping to change his necktie.

Some simplifying points need to be made. The first two categories are mostly different people: it is not usual to meet many active experts in germ warfare at a peaceful microbiological conference, or to find American and Russian missile engineers swapping ideas. The number of individuals in the latter two categories (high-ranking science advisers and Pugwashites) comprise a very small fraction of the scientific community. The normal professional exchanges of scientists provide the psychological base for Pugwash. On the other hand, familiarity with defence problems provides substance for Pugwash and, as a matter of policy, Pugwash wishes to involve people with influence in their national governments.

The most important conclusion from the evidence of events in the West since 1945 is that the bias of the majority of leading research workers is on the side of moderation in armaments, caution in international behaviour, and restless search for ways of abolishing war. When the politicians have taken the other course, and ordered another really big spurt in the arms race, they have usually done so either in plain defiance of scientific advice, or only with the support of advisers selected for their complaisant views. Perhaps too often other advisers have acquiesced in more moderate increments in armaments, that they found it politically inopportune to oppose. But at the big steps, like the use of the A-bomb, the development of the H-bomb and the introduction of the anti-ballistic missile, opposition has been robust. That this opposition is not without professional risks was shown as recently as 1969, when Franklin Long's appointment as director of the National Science Foundation was blocked because he had criticized the anti-ballistic missile programme in the USA.

For more than two decades, scientists have kept up constant pressure for disarmament, both in public and in the closed rooms of government. They have used rhetoric and ethics, analysis and inventiveness, to try to head off global disaster. They have forged their own quasi-diplomatic links between nations and set high standards of international conduct in their own activities. It is hard to see what more they could have done, unless more of the same. Research workers and technologists are responsible for the existence of new weapons, certainly; but then governments have been quick to milk scientists of weapons techniques, slow to heed their warnings of disaster or their proposals for peace. For the fact that weapons are now manufactured and deployed on a scale exceeding the wildest nightmares of 1945, and that the Earth is a most dangerous planet on which to live, the politicians and the military lobbies must carry most of the blame. There are other areas in the relationship of research and international affairs where the scientists may have been negligent, not here.

They continued to seek new forms of protest and action. In March 1969, faculty members of the Massachusetts Institute of Technology and other American universities stopped research for a token day, and held teach-ins about the continued abuse of science for war. In the following month a British Society for Social Responsibility in Science was launched; it was conceived after an earlier scientific meeting on biological and chemical warfare.

A political scientist and former member of the White House science staff, Eugene Skolnikoff (1967), reported on the scientists involved in the work of the US President's Science Advisory Committee and its panels: 'Certainly the role of these scientists in strengthening the nation's military posture and world power, at the same time they were actively seeking ways of bringing the arms race under control, cannot be denied. All that is certain is that the possibility of contradiction between these activities was a source of torment to many.'

CHAPTER 4

In the Automaton

IN 1961, two physicists at the General Electric Research Laboratory at Schenectady in New York State, Robert Walker and Paul Price, reported that if they dipped ordinary mica in hydrofluoric acid for a few minutes, rinsed it and looked at it under a microscope, they could see little tunnels. These were tracks originally ploughed millions of years ago by the fragments of uranium atoms in the mica that had undergone fission. They were now made discernible by etching. Other minerals showed the same 'fossil fission tracks'.

It was an interesting discovery, though hardly sensational by current standards. It turned out to have a range of consequences in scientific research and in utilitarian applications. This work, in which Walker and Price were joined by Robert Fleischer, began as a purely speculative research effort. Within a few years many uses had been found for the little tunnels. For example, by this technique, holes of very accurate size were made in plastic sheets, thus forming a fine sieve. Such a sieve separated cancer cells from among other cells, because of the cancer cell's relatively large size and rigidity. In the first instance it was used for filtering blood for medical research, but potentially it appeared to offer a means of detecting cancer cells in routine screening for blood cancer and for uterine and cervical cancer.

No administrator in Washington or any other capital, and no advisory group of scientists, however brilliant, could have anticipated that research on fission by an electrical company could lead to an improved means of detecting cervical cancer; or, for that matter, that the resulting technique might, through its use in dating, come to shed light on the evolution of man.

This example is recent, striking, but not at all exceptional;

there are plenty of others that show how confusing and anarchistic is the course of research. Uranium fission itself turned up quite unexpectedly and changed nuclear physics from an academic pursuit of learned gentlemen into a force that shook the earth. Research workers often refer to 'serendipity', which means finding something unexpectedly when looking for something else.

To give the impression that most research most of the time is lurching around in a drunkard's walk would be misleading. Major discoveries may be made by quite systematic interplay of experiment and theory, with no digressions. But even rectilinear motion homewards does not tell you what you will come across on the way or what the consequences for society will be. When Michael Faraday was playing with his magnet and coil of wire, he could not foresee that it would lead, by a quick series of steps, not only to electric power stations but also to radio, television, radar, radio astronomy and a new way of cooking frankfurters.

THE ILLUSION OF THE AUTOMATON

The ghost of Sir Francis Bacon must shiver o'nights, because he walks most frequently in Moscow. Sometimes, to be sure, he may be spotted on Pennsylvania Avenue, flitting into the Executive Office building, or by his beloved Thames on Millbank, summoned there by some self-confident planner of discovery and invention. You may run into him, disconcertingly, among the bungalows of New Delhi, or glimpse his reflection in the lake at Canberra. But a bleak Moscow day, with the citizens queuing in their boots and the huge television tower swaying in the gale, is the easiest time to see him, if not near the Gorky Street science office then out in the Academy territory along the Leninsky Prospekt. For Karl Marx, who was uncommonly hard to please, praised that Lord High Chancellor of England and a lot of people have never got over it.

'It is not surprising that Marx and Engels said that Bacon's philosophy occludes within itself the germs of a many-sided development,' wrote J. G. Crowther (1960). 'The similarity between the thought of Bacon and the modern formulators and operators of planned social development ally him more closely with them than with any other group of social philosophers and statesmen.'

Bacon did indeed ring a bell, as he claimed, to call the other wits of his time together. His bold and humane vision of a scientific state remains as inspiring as when he wrote it. Justice would be done, and the ghost laid to rest, if we could leave it at that. That he made an error should be a sidelight of no more consequence than, say, Isaac Newton's efforts in alchemy. How could Bacon know, after all, what scientific research would really be like?

In 1620, James I of England received from his Chancellor the first copy of a manifesto for the promotion of research, now generally known as *Novum Organum*. In half a lifetime of reflection, Bacon had come to envisage sciences and research organizations that did not then exist. He believed that their utility, based on 'a very diligent dissection and anatomy of the world', would be immense.

Had Bacon not been convicted the following year of taking bribes he might have had a chance to implement his ideas by official action. Bacon did not have to pay with his head, as he had let Essex and Raleigh pay with theirs. Instead he lived out his disgrace, to write *The New Atlantis*, in which he dressed his ideas in utopian form, in the description of the House of Salomon: 'The end of our Foundation is the knowledge of causes, and secret motions of things; and the enlarging of the bounds of the Human Empire, to the effecting of all things possible.' His imaginary island of Bensalem possessed, among other facilities, observatories and botanic gardens, pharmacological units and perfume-houses, means of aerial flight and meteorological laboratories where weather could be simulated.

Bacon brought together, imaginatively though imperfectly, the scientific ideas for the pursuit of natural knowledge then embryonic in Europe. Great experimenters like William Harvey and Galileo had leapt forward in ways that Bacon himself scarcely understood. But *The New Atlantis* reads as an astonishing anticipation of organized research in the scientific age that was to come. It does so even today, when many of the technologies Bacon asked for—from 'the mitigation of pain' to 'making new threads for apparel'—are matters of practical accomplishment. The men who founded the Royal Society of London explicitly acknowledged their debt to Bacon; so did the editors of the French *Encyclopédie*.

As Robert Hooke explained (1705), in Bacon's method of discovery the human intellect 'is continually to be assisted by some method or engine which shall be as a guide to regulate its actions, so as that it shall not be able to act amiss'. Although claims quite as bold as that are not to be found in Bacon's own writing, as a mild exaggeration Hooke's words will do. But Bacon's engine resembles the automaton with which Catherine the Great of Russia is reported to have played a game of chess—that is to say, it works very well, but only if there is a man concealed inside it.

Bacon's lesser error, whereby he overstated the role of fact-gathering and induction in making new discoveries, must be judged in the context of his time and his propaganda. But the lesser error led to the greater one, to the notion that research can be planned systematically—which resolves itself, on closer inspection, into the beliefs that anyone can be recruited to do significant research if he just goes about it correctly, that the course of a research project can be charted in advance, and that the beneficial applications of new knowledge will follow as surely as dawn after darkness. If you think that research and application are straightforward, that is where you finish up.

To be more precise, you finish up in the science office on Gorky Street, wondering why the Americans are doing so much better. The history of ideas is full of paradoxes, but none is of more significance for the planning of science and its uses than this: Bacon sought to throw out all dogma and all grand philosophy, and his view of science was adopted as dogma by the grandest of current philosophies. His fellow-countrymen, and the Americans, have followed Bacon much more faithfully in spirit by rejecting him in substance.

The paradox goes back to the differences of opinion between Bacon and René Descartes, three-and-a-half centuries ago. Bacon affirmed that knowledge was to come, not by deduction from preconceived ideas, but by induction from observations. Descartes, on the other hand, believed in the power of the individual human mind (in the first instance, his own) to make deductions about the natural world that could share the certainty of deductions in geometry. As it happened, neither man's method corresponded at all closely to the ways in which the new experimental philosophers came to succeed in practice. It was in a disorderly interplay of observation and thought, of induction and deduction, that they made progress; Galileo was a much better model for them.

Descartes' philosophy, being of a large-scale kind, is reflected strongly in the emergence of grand philosophies on the continent of Europe, bringing in their wake tough-minded political theories and the idea that research must be organized to serve the ends of the state. Descartes testified to the importance of thought, yet his method generated ideas that tended to suppress freedom of thought under overarching theory. The British empiricists, interacting closely with the natural science and political circumstances of their time, developed liberal ideas which accidentally created conditions in Britain and the USA wherein individual scientists and inventors could work freely.

The young Peter Kapitza (1922) wrote to his mother in Russia from the Cavendish Laboratory in Cambridge: 'England has produced the greatest physicists, and now I am beginning to understand the reason: the English school develops individuality on an extraordinary scale; it gives limitless room for a personality to show itself.'

FLASHES AND FUDGES

Even today, to many outsiders, scientific research appears to be a remorseless, humourless enterprise of computer-like minds dedicated to exposing the truth. The white-coated practitioner is seen methodically dissecting without pity, logically analyzing without imagination. This picture is not a caricature; it is just false.

Research is much less methodical than many other human activities, such as farming or tax-collecting. In reliance on logic, the lawyer or philosopher leaves the experimental scientist far behind. And 'truth' is much too big a word for the goals of research, which are merely explanations (plausible, aesthetically pleasing and mutually consistent for preference) of observed phenomena. The light that research casts on nature is itself a part of nature, the light of the human mind and not any extra-natural glare.

It is not even a 'computer-like mind'; no human being boasts such equipment. The great advances in knowledge come from hunter-like eyes that spot the unexpected, or from poet-like imagination that seeks rhymes and rhythms in the crude data of the senses. Systematic work may check the conclusions, fill the interstices of knowledge, or even help to transport the researcher into unexplored domains. But without the spark of cunning imagination men would have a surfeit of information and no understanding.

In Tunis in 1909, Charles Nicolle discovered that typhus was transmitted by lice—one of the crucial discoveries in the history of tropical medicine. The evidence on which it was based had been available to many doctors before Nicolle, in the fact that, while typhus was highly contagious in town and country, patients could be crowded together in a hospital without transmission from one to another and without much risk to the doctors and nurses caring for them. The interpretation came to Nicolle as he stepped over the body of an Arab, collapsed on the last steps leading to the hospital. 'When, a moment later, I entered the hospital, I had solved the problem,' Nicolle related later. 'I knew beyond all possible doubt that this was it.' The typhus was arrested at the reception office of the hospital, where the patient was stripped of his clothing, shaved and washed. The agent of infection was therefore something the patient carried on himself—nothing but a louse. Within two months Nicolle confirmed what he already knew, by experiments with lice and monkeys (Taton, 1965). There are plenty of such flashes in the history of science.

Nicolaus Otto, the German commercial traveller who invented the four-stroke-cycle internal combustion engine on which the automobile industry was founded, developed the underlying theory for his purpose by watching smoke from a factory chimney. The theory was quite wrong—but the machine worked. His ideas about 'stratified charge' had no bearing on modern automobile engines, although some engineers have subsequently applied the theory to attempts to design more efficient and less dirty engines (Bryant, 1967).

Some of the most important advances in theoretical physics—popularly thought to be the most rigorous of the sciences—were made by shameless 'fudges' that no schoolmaster would tolerate. To produce his fertile set of equations for electromagnetism (which among other things implied the existence of radio waves) James Clerk Maxwell had to invent a phenomenon that no one had observed, the so-called displacement current. In his original theory of the atom, which explained a huge area of observations, Niels Bohr mixed classical and quantum ideas with sublime inconsistency—but it worked. It was as if nature were in conspiracy with human hunches, and to hell with logic.

The immunologist Peter Medawar (1967) hears the working scientist 'telling stories', making tentative theories which must nevertheless be tested very scrupulously to find out if they are about real life. Medawar contrasts two conceptions of science: in one it is an imaginative and exploratory activity, with intuition the mainspring of every advance; in the other, it is above all else a critical and analytical activity, and imagination must at all times be under censorship. Both points of view can be correct, which may seem confusing for the outsider. As Medawar comments: 'There is no paradox here: it just so happens that what are usually thought of as two alternative and indeed competing accounts of *one* process of thought are in fact accounts of the *two* successive and complementary episodes of thought that occur in every advance of scientific understanding.'

A myth of the 1940s and 1950s was that inventions were no longer the product of individual brains but came from big research teams—a myth dispelled by John Jewkes (1958) and his colleagues in their inquiry into the sources of invention. If one would illustrate the continuing power of the individual imaginative inventor by a solitary example, a good one occurs in the case of heat transfer—the commonplace problem of keeping machinery and electronic equipment cool or, alternatively, of heating something. Efficiency of heat transfer is often crucial to performance and large numbers of physicists and engineers have devoted their careers to improving it, with significant but usually marginal gains. Metals, the best heat conductors available, were simply rather poor. Two Americans, Richard Gaugler and George Grover, independently hit upon the heat pipe, with sensationally better heat-transfer properties. It is a tube containing a metal wick saturated with a liquid that is volatile at the required temperature of operation; heating it at one end makes the liquid evaporate, the vapour carries the heat instantly to the cool end of the tube, where it condenses and returns by the wick to the hot end. Many engineering designs, and a good deal of industrial economics, are likely to be transformed by the invention.

IN SEARCH OF EXCELLENCE

A visiting journalist once asked the director of a famous research institute: 'How many scientists work in your laboratory?' The director reflected for a moment and then replied, 'Less than half'.

Manpower statistics and the allocation of funds to research and technology—these are the stuff of modern science policy-making. Yet they say very little about the quality, originality or importance of the work done. To talk about a thousand scientists is as bewildering as to talk about a thousand poets. When I see, in a statistical table, that in 1963–4 Sweden deployed 6340 qualified scientists and engineers, it conjures up no picture in my mind at all. I do not greatly err if I think only of a score of exceptional Swedish research workers and about the same number of outstanding technological developments that I know to have occurred in that country. The big battalions of science are necessary to keep academic and industrial establishments going, as a kind of intellectual cannon fodder, but whether one is looking for advances in understanding nature or money-making ideas for industry, it is the imagination and enterprise of individuals that counts. It is no exaggeration to visualize a scale of excellence—'outstanding', 'excellent', 'good', and 'mediocre'—such that the contribution to research or technology by an individual in one ranking may be ten times as great as that of an individual in the next ranking, so that one outstanding man may be worth a thousand mediocre ones.

Kapitza (1966b) put it more subtly: 'Between a genius like Newton and an ordinary man, the difference in ability is immense. . . . Suppose you take a Newton and add to him ten ordinary scientists, you have the ability equal to two Newtons. You add to this institute one hundred people more, but the ability would be only three Newtons: the scientific effectiveness grows only in proportion to the logarithm of the number of people engaged. So, if you miss a Newton, that is equivalent to missing a large number of people of ordinary talent.'

A scheme for trying to spot budding scientists in the USSR, by nation-wide 'Olympiads', starts with problems published in the newspapers or announced on television. Those schoolchildren who send in the best answers are given more difficult problems at regional centres, and the best of those face a final, very difficult problem in Moscow.

The youngsters so selected formerly went to leading universities, but Kapitza sought to bring them into research institutes, where they could work with the ablest scientists and the best equipment. After some years of difficulty with this scheme it suddenly began to work properly, in the mid-1960s, when the first students under the scheme became teachers. Thus, there was growing awareness among leading Soviet scientists and educators of the role of individual excellence in scientific research.

TESTIMONY OF PURITY

Every 10th of December, on the deathday of the inventor of dynamite, a solemn ritual takes place in Stockholm. Outside the Concert Hall the snow sits like so many bonnets on the figures of Carl Milles' *Orpheus* and the temperature is well below freezing. Inside, distinguished men from various countries bow to the Royal Family of Sweden and take their seats on the stage. Everyone is in evening dress although it is mid-afternoon. There are speeches in Swedish explaining the achievements of the men on the stage; white-robed girls from Stockholm University sing in the gallery. One by one, the men come down from the stage to receive Nobel medals and cheques. For most of the scientists concerned, walking down those steps to where the King of Sweden is standing to greet them is the greatest moment in their careers.

This curious midwinter festival serves like ancient magic to ensure that spring will return and scientific work in the future will be as fruitful as in the past. The Nobel awards, made with meticulous judgement by committees of Swedish scientists, give researchers in all countries a standard of excellence to which they must aspire. That remains so, even when Nobel's categories are plainly unfair to scientists in certain fields, and when in any case there are not enough prizes to go round. The way the awards impress both scientists and general public confirms that research remains wholly dependent on the talents and dedication of outstanding men.

There are other prizes—for literature and, when it does not seem too ironical, for peace. But this is primarily a scientific ceremony, and mostly for 'pure' scientists at that. Even the medical prizes are awarded for work with a big element of discovery. Few of the practical advances in 'useful' science and technology in this century have taken their orginators to the Stockholm Concert Hall. Nobel's will referred to benefits to mankind, but the juries have leaned heavily towards discoveries that were primarily important for the enlargement of knowledge.

Indeed, the Nobel physics and chemistry awards are clear public evidence that the distinction between 'pure' and applied science is not, as some would have it, a fiction of academic snobs. It is just as well that the distinction does have validity, even though one would be unwise to attempt a precise demarcation. If talk of 'pure' and applied research is forbidden, it is hard to make sense of the morality of science; nor can one hope to find ways of regulating the applications of knowledge without compromising the pursuit of knowledge.

The greatest single bar to clear thinking about social interactions or planning of 'science' is the word itself. Never mind what it ought to mean; even professionals use it in a very loose way to cover, variously:

a body of knowledge: 'I'm learning science at school'
a process of acquiring knowledge: 'Science will give us the answers'
the application of knowledge: 'Science has changed our lives'.

Of these three, the *process* is really the central idea, but then to define the process is not straightforward. Although philosophers have tried to codify 'the scientific method' (Bacon's engine) there is really no such thing; or else there are dozens of scientific methods and anyone is free to invent a new one. What matter are the attitudes of the people concerned, and the social system whereby they check up on one another's results and theories. That was what Desmond Bernal implied when he said: 'Science is what scientists do when they are working.' A scientist, for present purposes, is someone who takes part in a particular social system, essentially an ethical system—though a very practical one.

The scientific process is of proved effectiveness in some fields of inquiry, and less well-established in others. 'Science' has a narrower, and therefore more descriptive, scope in common English usage than have its equivalents in other languages, where *Wissenschaft* and *nauk* embrace history and philosophy—and so do corresponding institutions. In English usage, the core of science is the experimental natural sciences—historically, physics, chemistry and biology.

Ambiguity remains about whether 'science' includes or excludes its own applications. Medical research (or human biology) is a venerable science—but does that make medicine a science? Engineering and the other technologies are becoming increasingly dependent on scientific research; there is no clear dividing line, so it is accurate simultaneously to distinguish and link them as 'science-and-technology'. Journalists and other non-scientists, who abhor mouthfuls, abbreviate the expression once more to 'science', which then embraces technology. As the engineer complained about the reporting of spaceflight: 'When it works, it's a scientific achievement; when it doesn't, it's an engineering fault.'

The ellipsis is more dangerous than that. As Hyman Rickover (1965) pointed out: '. . . technology cannot claim the authority of science. It is properly a subject of debate, not alone by experts but by the public as well. It has proved anything but infallibly beneficial.'

The usually unfaultable Medawar (1967) entangled himself in trying to deny the difference between pure and applied research. His argument was that purity was not regarded as a virtue when scientists were judging their colleagues' discoveries and interpretations. Instead, 'foremost is their explanatory value—their rank in the grand hierarchy of explanations and their power to establish new pedigrees of research and reasoning.' Clarifying power and the feat of originality involved in the research are further considerations. 'But purity, as such, is nowhere. Nor is usefulness, which has its own scale of valuation and its own rewards.'

Yet that is a very testimony of purity, as it is commonly understood. Nobody suggests that an engineer cannot make a contribution to knowledge; but by the standards of 'pure' science, its utility is of no significance, for or against. An academic scientist can afford to rely on the judgements cited by

Medawar, but an industrial research director would be out of a job if his men produced splendid additions to the 'grand hierarchy of explanations' but nothing of use to the company. In trying to decry the snobbery associated with pure science, Medawar in effect merely invites the applied scientists to join the club of 'pure' scientists, whenever it is opportune to do so.

THE PRACTICAL ETHICS AT THE HEART OF SCIENCE

On the day that Yuri Gagarin made the first flight by man in space in April 1961, the world's space scientists were in conference in Florence. When the news came they were discussing 'special events', but that referred to the big flares on the Sun observed in instruments in unmanned satellites. It is still astonishing to recall that the scientific session went on without interruption that morning, so that many of the scientists were unaware of what had happened; and it was considered unnecessary to take formal note of this special event. Hendrik van de Hulst, president of the international Committee on Space Research, produced some wine for Anatoly Blagonravov, head of the Soviet delegation, and the mayor of Florence came bustling to bring congratulations; next day, Alla Massevich was to be mobbed by excited Florentines after giving a public lecture in the Palazzo Vecchio. But the scientific assembly was strangely unmoved.

This pattern of behaviour illustrates an aloofness among many scientists that is hard to explain to non-scientists but which affects the relationship between research and its practical applications. Because of his expert knowledge, the scientist tends to discount in advance even the epoch-making achievements of his colleagues and his sophistication may make it hard for him to share the sense of wonder or surprise experienced by the general public. Moreover he is inclined to take a severely disciplined view of the purposes, methods and results.

'Pure' research is the heart of science. It operates according to internal rules, mystifying to the outsider, which are responsible for quality control, for ensuring the reliability of the knowledge produced by research. American sociologists, particularly Robert Merton, Bernard Barber and Norman Storer, have sought to identify the standards of behaviour that scientists set for themselves. They turn, as Storer (1966) concludes, on the scientists' wish for 'creativity'. That is the commodity in which the scientific community deals, like the 'goods' of the economic system, or the 'power' of the political system. The terms of exchange in the 'social system of science' are essentially, as Storer puts it, 'the offer of one's creative work, in return for response, competent evaluation, from others'. Other creative workers such as engineers and writers have been unable to evolve quite the same interchange with one another—largely because the 'audience' for their work is usually non-expert.

This analysis helps to validate the idea of 'pure' science and also explains some oddities in behaviour. An 'ivory tower' attitude is positively encouraged by the requirement that research workers should aim to impress their fellow scientists rather than the outside world. Scientists are often slower than laymen to accept the correctness of new discoveries, but the opposition to revolutionary thinkers that seems to mar the history of science really illustrates a conservative principle that operates against the crackpot; for every genius who has to wait for the old men to die before his ideas are accepted, hundreds of silly ideas have been prevented from contaminating the body of scientific knowledge.

When Merton encountered a direct resistance from scientists to the sociological study of 'multiple discoveries' he ascribed it to 'a resultant of intense forces pressing for public recognition of scientific accomplishments that are held in check by countervailing forces, inherent in the social role of scientists, which press for modest acknowledgement of limitations, if not for downright humility' (quoted by Storer).

For peculiar but important reasons of their own, when research workers come publicly to report and discuss their results, they do so very coolly. They deliberately distort the narrative so that it sounds as if it all went logically and methodically. The reasons have to do with the way prior work must be acknowledged and the new work is to be evaluated and exposed to checking or refutation. The consequence is that in journals, text-books and the classroom, the scientists show not the man, but the Automaton.

'I have never seen Francis Crick in a modest mood.' This exception to the rule of public statements by scientists caused a sensation. James Watson, who at Cambridge in 1953 was co-discoverer with Francis Crick of the structure of DNA, the genetic material, gave an unprecedented personal account of how a discovery happened (Watson, 1968). 'I explained how I was racing Peter's father [Linus Pauling] for the Nobel Prize.' It was a tale of bumbling pursuit of the truth, foolish errors by great men, intense personal competition and squabbling; of how the young American research student's thoughts alternated from girls to genes, depending on the accessibility of the former; and of how Crick and he found the answer by fiddling with molecular models.

Watson's closest colleagues of those days opposed publication and his own university press dropped the book on the instructions of the Harvard Corporation. The editor of the British journal *Nature* tried a dozen distinguished molecular biologists in an effort to find a reviewer for the book and commented: 'It would probably have been easier to obtain a molecular biologist's opinion on a piece of out-and-out pornography' (Maddox, 1968).

Although everyone knew that scientists were not razor-minded saints, they had to pretend they were, otherwise the social system of science would not work. The ethics at the heart of science is, in short, not so much a set of rules as a system, mediated by institutions like societies and journals, for sorting the sound from the unsound, the brilliant from the pedestrian. It is extremely flexible as to the content and methodology of the research being tested and accepted; it can cope with current speculations about quasars and quarks as it did with the observations of Hooke and Halley. The system works superbly well, with much greater speed and general good humour than any other modern social system. Telegrams, letters, visits, seminars, conferences and preprints of papers carry the news from laboratory to laboratory among the sub-systems of specialists. A major discovery may be confirmed and extended in half a dozen countries within a few days of its announcement.

Yet in public debate the scientists themselves are relying increasingly on the likelihood of ultimate utility for their discoveries as a justification for 'pure' research, and even well-meaning and well-informed political leaders take the same line. What else was implied by the following remarks of John Kennedy (1963) to the National Academy of Sciences?

> 'Even though some of your experiments may not bring fruition right away, I hope that they will be carried out immediately. It reminds us of what the great French Marshal Lyautey once said to his gardener: "Plant a tree tomorrow." And the gardener said, "It won't bear fruit for a hundred years." "In that case," Lyautey said to the gardener, "Plant it this afternoon." That is how I feel about your work.'

Observe the belief in the eventual pay-off, the view that 'pure' research is just a slow kind of applied research. To say that much of present-day 'pure' research will turn out to be useful in the long run is a reasonable conjecture, based on past experience. Most scientists certainly believe it. Where apparently mild but potentially devastating slippage begins is in the implication that we pay for 'pure' research today *in order* to use the results in the future. There are implied obligations: on science to be useful, on society to use it, wherever it may lead.

CHAPTER 5

Occident Express

PEASE POTTAGE hill is not dramatically steep but, climbing up to the central forest ridge of the English Weald, it is sufficiently long to test severely the motors of antique cars on their way from London to Brighton. An annual run of 'old crocks' commemorates the Locomotives on Highways legislation of 1896, which abolished the man with the red flag. When early 'locomotives' ventured on the roads in Britain, the law required that each should be manned by a crew of three, one of whom had to walk in front with the flag. The maximum speed was 4 mph, and 2 mph in built-up areas.

Since the liberation of the car and its driver, more than a quarter of a million people have been killed on British roads. By 1968 the cost of road accidents was being quoted by the Minister of Transport at £250 million a year. The automobile forced expensive and often ugly reconstruction, first of the nation's highways and then of its cities. Noise, fumes, congestion, parking meters and Breathalysers were the outcome of seventy years of this technology. If the parliamentarians had foreseen the consequences, they might have preferred to keep the man with the flag foot-slogging in front, to this day. He would provide reserve manpower for pushing the vehicle up Pease Pottage hill.

Did the legislators really have that option, or were they bound to yield to the demands of the deadly technology? A modern nation without fast, private road vehicles is hard to visualize unless one also imagines what other means of public transport it might now have. Even then, to defend the idea that the automobile need not have come is not a very easy piece of historical conjecture. The impression is strong that science, technology and innovation sweep remorselessly on, outside human control.

It is nevertheless an illusion—more precisely a set of related illusions. The Illusion of the Automaton we encountered in the last chapter. Four others, which I call the Book, the Turnpike, the Marriage and the Railway are added in the table, and discussed in the following pages.

THE ILLUSION OF THE BOOK

In retrospect, the discovery of scientific knowledge has so strong and dramatic a story-line, with such appearances of logic and necessity, that its course of development seems almost inevitable. More than that, men have an impression that the Book of future knowledge is already written by nature, which made both the external world and the inquiring human brain. As we turn the pages, the story unfolds. The most that an ambitious man can aspire to do, it appears, is to guess what is coming next, which may entitle him to write his name in the margin of history.

Physicists became interested in electricity, discovered that atoms were decomposable, resolved the atomic nucleus into nucleons, found that nucleons too were decomposable . . . and by the late 1960s everyone was looking for the quarks, the postulated master particles at a more fundamental level. So it goes on.

The Illusion of the Book is not merely convincing but productive. For planning research itself, it is as good a model of scientific progress as any. Research leaders consciously try to guess 'what is coming next', in choosing areas ready for research, and they often succeed. There are fashions in science, as large groups of researchers switch to the same new line, exploiting a recent discovery or technique and striving for the next. To engage in scientific research at all, one must believe that nature has a stockpile of phenomena and laws accessible to the human brain, if only it will read on patiently. Nature has not disappointed it yet.

The sequence of discoveries in the Book is determined at least in part by the fact that men can take ideas in only gradually, step by step. When the plot has reached a certain point—when techniques and theories are ripe, and nature has something lurking just beyond existing knowledge—someone will turn the page. In fact, two or more people are likely to do

---

IV. A CLUTCH OF ILLUSIONS

Area of policy	*Illusion*	*Nature of illusion*	*Reality*
Research and manpower planning	'AUTOMATON'	There is a quasi-mechanical scientific method that almost anyone can master	There is a clever man inside the 'automaton'
Research planning	'BOOK'	The course and sequence of discovery of new knowledge are inevitable	The 'plot' can be altered
Planning for innovation	'TURNPIKE'	There is a clear road from research to application	The 'road' is very crooked
Management of innovation	'MARRIAGE'	New knowledge is automatically applied whenever practicable	Application depends on the enterprise of individuals
Control of the uses of science	'RAILWAY'	The course of technological civilization is inevitable	There are plenty of branching points on the 'track'

Logical connections

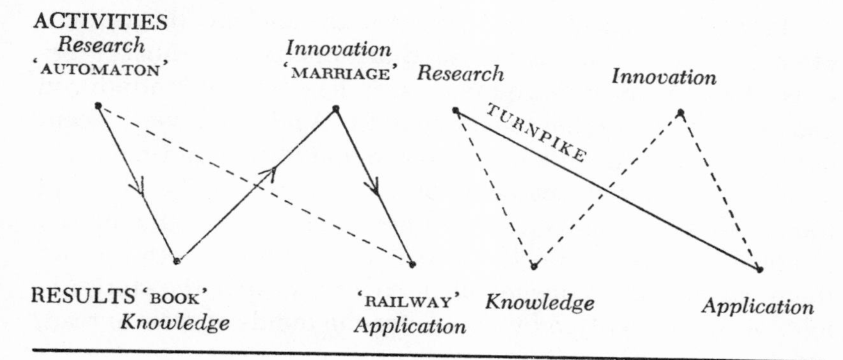

so, quite independently, yet almost simultaneously. Wrangles ensue about who did it first and who can write his name. Indeed such 'multiple discoveries' and the continual race for priority between scientists are among the strongest evidence for the Book; so is the very plausible idea that 'if Otto Hahn hadn't discovered nuclear fission someone else would have done'.

Yet the characters in the story are men, and their preoccupations or whims must influence the sequence and timing of the phases of the story. Suppose that something had distracted the leading physicists and radiochemists, including Hahn, from nuclear studies in the mid-1930s. For example, there was no sequential reason to stop someone then inventing the laser, the astonishing new light source that in fact turned up in 1960. Postpone nuclear fission—and there might have been no Hiroshima A-bomb, no nuclear arms race, nor yet any nuclear power stations or political enthusiasm for further nuclear research. With nuclear fission so close to discovery, it could hardly have remained unnoticed for very long. Someone might today be dipping mica in hydrofluoric acid and wondering what on earth could have made those little tunnels. Even so, development of research, as well as of society, would have been greatly affected by such a reversal of events. The death ray would have been the big obsession of the second World War (probably futilely); today there would be far fewer nuclear physicists and far more specialists in optics. The story would have been quite different. Not for a hundred, or perhaps a thousand years, would this inverted chronology cease to be significant.

The social connections of research are ancient, diverse and often obvious. Lycurgus banished arithmetic from Sparta, because it was democratic and popular in its effect, and introduced geometry, better suited to an oligarchy. Again, in a very recent example, practical work on radio communication and radar led accidentally to the detection of radio noise in the sky and thence to the quasars and pulsars, great discoveries in the physical sciences in the 1960s. Is it then possible both (1) that there is a natural sequence of discovery, so that the story unfolds as a book written by nature for the minds of men to read, and (2) that at the same time social circumstances, interests and needs influence the story? Not unless one supposes that social developments are also inevitable. Nowadays, social developments are to a great extent influenced by the applications of science: are those applications predetermined too?

TURNPIKE AND MARRIAGE

Around the northern circumference of Greater Boston, the ring road Route 128 is world-famous—at least among those who fret about the encouragement of new technology—as the site of an amazing range of science-based industry and industrial research companies. Polaroid, Sylvania, Avco, High Voltage Engineering . . . such are the signs you see along it. Less well publicized, but of greater significance for theorists of innovation, is the Concord Turnpike, that links Route 128 with Cambridge, and thus with Harvard and the Massachusetts Institute of Technology. The establishments of Route 128 are like extramural activities of those universities, being commonly either founded by faculty members with money-making ideas or drawn there by the wish to recruit able scientists and engineers, who in turn will make it easier to win fat government contracts. At MIT, where students have formal instruction in how to found their own companies, visitors from the firms up the road mingle indistinguishably with the faculty.

The Concord Turnpike seems to give physical embodiment to a simple-minded belief that there is a clear road leading from research to application of the results of research—and on to the bank with profits. The Illusion of the Turnpike survives because a skilled driver can often make the journey deceptively fast and easy. But in reality it is a crooked and hazardous road, as likely to lead to the bankruptcy court as to the bank. And the notion of the Turnpike neglects the subtle process that turns a discovery or invention into an innovation. In the case of Route 128 and the nearby universities, it is the interchange of men and their ideas that counts, not the mere propinquity of laboratory and factory. That it is now unwise to segregate research, innovation and exploitation in commercial thinking does not alter the fact that the Illusion of the Turnpike linking them has for long confounded technologists, managers and investors. So has the closely related Illusion of the Marriage.

Research and application seem to be faithful partners. To the casual eye, discovery mates with purpose at every opportunity and populates the social scene with innovations.

Yet again, the illusion is almost believable. The electron is discovered and, after a minimal gestation time, electronics is born. The navy wants a way of detecting submarines so the scientists produce sonar. Radiochemists detect nuclear fission . . . bang. The doctors must try to stop polio, so the medical researchers devise vaccines. Physicists invent the laser and, in no time at all, eye surgeons are using it to treat detached retinas and forensic experts are considering it as an aid in identifying fingerprints. What harmony! What fertility!

The Illusion of the Marriage prevails if you regard a common event as a general, automatic rule. Publicity and history attend the innovations that happened, in preference to those that were stillborn or never conceived. If the link between research and application were reliable, there would be no cause for anguish among science policymakers about the inefficiency of the process of innovation in their respective countries. What need for all those exhortations to industrialists to look to science, or for researchers to come out of their ivory towers, if the Marriage were real?

A clearer view of what happens came in studies for the National Academy of Sciences in Washington and the Organization for Economic Co-operation and Development in Paris. Both were based on the realities in the USA, the innovating country par excellence. The NAS panel led by Harvey Brooks (1967) reported to the US Congress:

> 'New ideas may come from almost anywhere: from a basic scientific discovery, from recognition of a market need, from a management decision, from an obscure inventor, from the technical literature, or from a casual conversation at a pro-

fessional meeting. A key factor in successful applied research is the achievement of voluntary co-operative effort among scientists and technologists with diverse backgrounds—the association of people who became sold on a new idea.... The best atmosphere for applied laboratories is characterized by internal freedom under strong leadership. Success in applied research is seldom achieved by authoritarian methods.'

Next, Joseph Ben-David reporting to the OECD (1968):

'Trying to make science more useful for practice, in the common-sense way, by preferring research projects with obvious practical applications, will be self-defeating.'

With this remark the Israeli sociologist exposed the central paradox of science policymaking more clearly than ever before. His comparison of performance in the USA and Europe led him to emphasize American 'entrepreneurship' as the means of technological advancement. In this very important passage, Ben-David went on to use a rather obscure analogy from physics to describe the process. I prefer a simpler sexual metaphor, which converts the Illusion of the Marriage into a more realistic picture of promiscuity.

Scientific discoveries are female and practical purposes male; the offspring are innovations. Just as a woman may find it hard, in retrospect, to imagine herself married to a different man and therefore encourages romantic propaganda about marriages being made in heaven, so it is difficult to visualize that the discovery of uranium fission, for example, might have entered into liaison first with something other than the practical purpose of military explosives. Hence the tendency to assume that such exploitation was the natural, pre-ordained course of events.

In the real world of technology, as of sexual behaviour, chance encounters occur between discovery and purpose; frequently there is flirtation, sometimes there is coupling, occasionally the coupling is fruitful. Some discoveries, electricity for example, serve as common whores. Some purposes, most notably the military, are very wolfish, always on the look-out for virgin discoveries. Other purposes, including basic ones like education and urban planning, have been relatively blind to discovery until very recently.

The policymaker who wishes to increase the birth rate of technological innovations must promote promiscuity; he must make sure that there are plenty of discoveries and that institutions are well adapted to bringing about the necessary chance encounters. Human individuals enter the process in a dramatic way as technological entrepreneurs who respond both to the disclosures of nature and to the urges of society. But they are non-conformist individuals with passions and weaknesses; their imaginations largely determine in which direction creative enterprise operates. Just as we found a man skulking in the Automaton of research, so now we find another at the very hinge on which the Illusion of the Marriage relies.

The NAS panel already cited also reported that 'Applied science probably needs more heroes'. At least people have a stereotype of the inventor (mad or otherwise); for the new style of American entrepreneur they may need a prototype. Hyman Rickover may do, as the unconventional admiral who fought for and then masterminded the US Navy's nuclear submarine programme. He is, incidentally but not surprisingly, a man greatly concerned about the impact of technology on human and natural life.

Intellectual promiscuity is a characteristic of American science-based industry. Research workers and salesmen talk to one another. Senior men are accustomed to the idea that an innovation revolutionizing an industry may originate from an obscure and junior employee. Huge budgets go to maintaining the flow of information and ideas, whether by paper or word of mouth. The inventor and the technological entrepreneur trigger the innovating process, but there remains a succession of steps involving, for example, technical evaluation, decisions on research investment, study of cash flow, decisions on production—with all the time the need for a shrewd eye on costs and prices, on the market, on the future, on direct competitors and on rival technologies. The Americans know at least as many reasons as anyone else for *not* developing a new idea, but they are deterred less readily. Their managers and investors are innovation-minded, in a capital market where what matters is not the dividend but the growth in stock value.

By the time a positive decision is taken in favour of an innovation, many questions have been asked, agonizing analyses of many kinds have been ventured in order to guess whether the new product can be sold profitably. There will have been anxious thought about snags that could put off the customers or invite suppression by the authorities. But it is unlikely that enough time will have been devoted to those other simple questions: What kind of world will this innovation be making? And will it be a better world?

THE ILLUSION OF THE RAILWAY

The portrait of a Westerner in the late twentieth century is a picture of technologies among which the outline of a human being remains discernible. The body and brain of this favoured citizen of Technopolis are indistinguishable from those of his forefathers of the Old Stone Age who lived by hunting, except that he is in some important respects healthier, in others less so —and he is better informed. He takes vaccines and pills to ward off ills, and other pills or spirits or filtered smoke to console him. His sexual activities are usually frustrated in their biological purpose by contraceptives; but when he has children he brings them up with due respect for the vulnerability of their unconscious minds. He drinks filtered and fluorinated water, eats packaged foods prepared from genetically selected plants and animals, and clothes himself in the products of chemical works.

He augments his locomotory powers with a mass-produced wheeled vehicle driven by refined spirits of mineral oil. His vision is extended by a box of electronics that enables him, simultaneously with millions of others, to witness—in colour if he wishes—events or recordings of events taking place thousands of miles away. If he wants to speak to a friend or colleague he can do so without moving from his chair, by telephone. If he wants music, the recording industry provides it; if he wants false teeth they are readily available. He lives and works by the clock, itself synchronized to atomic clocks that are more reliable even than the apparent motions of the stars.

In his occupation the Westerner feels himself a part of an economic apparatus dedicated to the generation and distribution of goods and services; policy is subordinate to the need to keep the wheels of industry turning and multiplying. His dreams parade before him women beautified by cosmetic chemistry, cars bigger than the one he has, jet travel to romantic islands where he will still be able to buy iced Coke. His nightmares are less frequently of monsters or nuclear catastrophe than of personal failure to keep pace with the social machine and its underlying technologies.

In such circumstances more people lead richer and longer lives than ever before. But no aspect whatever of ordinary human life in the technologically advanced countries of the West is unaffected by scientific knowledge and its uses; these grow, laying rails to a stranger future.

The apparent inability of the human species to make discriminating use of science induces a sense of helplessness. The impression is that we are passengers in a runaway train, hurtling faster and faster along a track. Some passengers try pulling the alarm signal, but nothing happens; there seems to be no one in the driving cab. We do not know where the track eventually leads, but the direction is apparent for the time being. It carries us towards new weapons, new industries, new mechanical and electronic products, new drugs, new pollutants, new sources of noise; it takes us away from nature and the simple life. Other trains on the line represent the different nations; some are far behind, but they are all going the same way. There seems to be no escape, no diversion open. 'You can't stop progress,' people say.

'Everything that is theoretically possible is bound to be realized in practice however great the technical difficulties may be,' declared Anatoly Blagonravov (1967), a leader of the Soviet space programme. The obsession of the 'progressive' politicians, technologists and industrialists is to move along the track as quickly as possible. Never mind where it is going, the important thing is 'to keep a lead', if you have it, to 'narrow the gap' if you don't. The military want new weapons, investors and wage-earners want economic growth, engineers want new challenges. Everyone wants what Americans already have; Americans want still more. The route is, in that sense, definitely westward.

Experts make forecasts about technological innovations in the next twenty years. If, as is more likely than not, many of the forecasts are proved correct, will that not testify to the Railway as a social fact, rather than an illusion? The forecasters generally assume that present political urges in the world will continue, though without catastrophe. A change of direction of these urges can switch the train to another route. Aha, the determinist will say, but these political urges will not change direction: would but can't, can but won't—there is no practical distinction. Yet by technical action, with or without political motivation, the route is also variable. A safe prediction is that several *unpredictable* discoveries and innovations of great importance will alter the human scene during the next few decades.

These novelties will come, for the most part, by quirky advances in research and by conscious efforts of scientists and engineers to do something useful. The discoveries will depend on what nature has to tell us, and on where the interests of research workers, inventors and technological entrepreneurs lead them. Can there then be an inescapable course of technical advance, even though that course is dependent not only on political decision (which may admittedly be derelict) but also on the activities of individual discoverers and innovators? No, it is not possible—unless you also believe that the course of research is naturally and socially predetermined.

It turns out, in short, that, just as the Book depended on the Railway, so does the Railway depend on the Book. For both to be reality, rather than illusion, the course of research must be fully compatible with the tendency of the technological society and that tendency must reflect and exploit, rather efficiently, the results of past, present and future research. The Marriage of research and application is crucial to the survival of the deterministic ideas. No technological problem which research can help to solve must be for long neglected; no innovation suggested by research must be overlooked. If it is possible to neglect an innovation by oversight, it is possible also to reject it by choice.

But there is no Marriage; so neither is there a Book or a Railway. Science, innovation and application are human activities subject to human decision.

It would have been much easier to dispel all the illusions at one blow, but I make no apology for labouring at them. Policymakers find it convenient to believe them all, severally, or at least act as if they do so. Others who do not wish to believe them, and the fatalism they imply, nevertheless despair of the possibility of controlling the course of our technological advance. To show them as an interlocking set (see Table IV) that stands or falls together, may help us to be rid of them for ever. And showing the men at the weak hinges reveals the points where deliberate promotion and control of innovation must operate.

BIRTH CONTROL FOR INNOVATIONS

The single puff, which should blow all fatalistic illusions away, is this: no country, however rich and determined, can afford any longer to pursue all the technological opportunities open to it. There are simply too many of them, and in any case they multiply faster than they can be followed up. Even if men do not modify their general nationalistic and materialistic ambitions, governments and technologists are still going to have to make arbitrary choices and determine priorities. To suppose that opinionated politicians, eager innovators and independent researchers are all mere puppets in the hands of fate as they help to make those choices calls for reserves of pessimism that few can possess.

Regulation of the uses of science and technology is certainly possible. Many are preferentially encouraged by governments with research grants and development contracts. A decision *not* to support a particular project is in itself a measure of choice and control, and that decision may be wise or negligent. Other technological developments are subject to restrictions, like noise-control requirements for aircraft operations or the laws governing weather-control experiments in various American states. Or regulation may be by veto. It is not surprising, for example, that few have heard of the project for abolishing darkness: it was not allowed to happen.

The space science board of the US National Academy of Sciences crushed the idea of putting a huge mirror in orbit about the Earth. The mirror, nearly half a mile across, was supposed to act as an artificial moon to illuminate portions of the dark side of the Earth. The National Aeronautics and Space Administration had contracted with five companies for feasibility studies. There was military interest, too, in the possible use of such a mirror for the night war against the Viet Cong. But the space science board reported in 1967 that it could see no scientific value in a satellite reflector system commensurate with the cost and nuisance to science of such a system. No such satellite should be considered unless it could be destroyed by signal from the ground or before detailed studies of its effect on ecology, biology and astronomy had been conducted and made public. Donald Hornig, presidential science adviser, thereupon said that the US government no longer planned such a project, and he promised that, if interest in the concept should revive in the future, the academy would be consulted again.

If that sounds too much like the scientists protecting their own interests, consider how Americans were defended against thalidomide. For four years, from 1957, *alpha*-phthalimide-glutarimide, discovered in Germany, was widely marketed in a world hungry for a safe sedative. But it was not safe and among its side-effects was a risk to pregnant women taking the drug. Many of their children were born without limbs before the tragedy was noticed, and stemmed. Animal experiments that would have shown this outcome and other side-effects had been skimped. In stricken countries, justice was retrospective.

The thalidomide experience did not happen in the USA because experts of the Food and Drug Administration, hardened fighters against quack medicines and dangerous drugs, would not let the William S. Merrell Company market the drug. They were aware of hints of trouble appearing in foreign journals, and medical suspicions that arose from them. In the Senate, Estes Kefauver (1962) proposed that Frances Kelsey, the FDA doctor chiefly responsible for the ruling, should be given a federal gold medal. Do not deduce that FDA was always a second Daniel. At that time Kefauver was struggling to strengthen its powers against the big drug houses; not until later were the regulations altered to make it necessary to check that drugs were beneficial as well as harmless. Nevertheless, in America, in the biggest pharmaceutical scandal of the post-war era, the regulations worked.

To control the outcome of research or invention is impossible, but that need not matter at all; what counts is the use we make of the results. Birth control for innovations is possible whenever we decide to employ it; family planning is perhaps a better term, because more positive. And therein lies the means, stronger than any other we have, even including idealistic politics, for making whatever kind of world we prefer. We are not helpless passengers, merely negligent engineers.

THE ETHICAL GAP

The task of applied research and technology is to serve purposes determined mainly from the outside world. The scientific merits of the work are incidental; what matters is utility. The institutionalized ethical system of 'pure' research provided a landmark (see the previous chapter) but for applied research and technology there is no equivalent system.

To a first approximation, the ethics of the application of science should mirror the ethics of society at large. It would seem intolerable if it did not. Research workers and engineers should not exploit their privileged position to steer society. They ought not to wish a new product on the rest of us only because it is scientifically and technically sweet to them; nor should they, beyond the limits of individual disposition or conscience, refuse to pursue particular developments that their community requires—whether these be philanthropic or profitable, medical or military.

Some who fear the onrush of technology take apparently contradictory positions, holding both:

> that 'the scientists' are in an ivory tower, neglecting to accept responsibility for their discoveries or to guide society in the scientific age;
>
> and
>
> that 'the scientists' have taken over and are disregarding the 'real' needs of society.

Behind the foggy thinking represented here, lurks the central and ancient problem of expertise in society: simultaneously we want, and do not want, the experts to tell us what to do with their expertise. (They, simultaneously, both resent instructions and crave for them.) We want them 'on tap not on top' but we still want to feel free to blame them if our decisions are disastrous.

The technologists and the users of the applications of science (including governments and public) are operating in an ethical gap, between the ethics of 'pure' science on the one hand, and the conventional ethics of social and political life on the other. The first is unconcerned with application. The second is not attuned to scientific issues; it cannot say promptly whether psychedelic drugs are a liberation like alcohol or a blight like morphine, whether ocean research should be a unifying force or a cause of war, or whether eugenics is a good idea. The institutions and people are not well adapted to dealing with such questions.

What about the idea that science itself may show us what to do with science? Men of science have often looked for a new, secular basis for ethics. Some say that the ethical system of 'pure' research for ensuring intellectual freedom and honesty provide a model for human ethics in general. 'If all the world conducted its affairs, and interacted and reasoned as we do,' they say, in effect, 'all would be well'. Yet it is probably undesirable to try to imitate, for example, the scientists' professional quest for consensus; lacking the benefit of reproducible experiments, social affairs must remain controversial.

Biology is a rich seam for the would-be quarrymen of new ethical systems. Typically we are supposed to spot the tendency of human evolution, and to judge as 'good' the acts that conform with the tendency. C. H. Waddington, for example, argues that the human code of evolution, which depends on transmission of information, requires as a functional part of its mechanism 'something which must have many of the characteristics of ethical belief'. A biological analogy for ethics which has recommended itself to Alvin Weinberg is the tendency of a living organism to produce order out of its disorderly surroundings; in physicists' language, the organism reduces its total entropy. Is whatever increases our order therefore good? Weinberg's idea is that 'merit is to be judged by the degree to which the activity, scientific or human, contributes to the unity and illumination, and ultimately to the harmony, of the many activities with which it interacts'. It looks too general a concept to be of much use as an ethical guide to action; it is also antagonistic to the merits of diversity.

The possible development of a systematic science of ethics—beginning with a description of men's morals, then comparing the morals of different people at different places at different times, and finally striving to become explanatory—was pondered by Henri Poincaré (1913). But he was sceptical about its practical value. 'It can be no more a substitute for morality than a treatise on the physiology of digestion can be a substitute for a good dinner.'

Bertrand Russell (1959) simply dismissed any hope of finding a secure basis for ethics in science, sticking firmly to the view that science was concerned with means—how best to reach certain ends—and could not tell a man which ends he should pursue. Perhaps the optimists who still seek an ethical system in science will confound Russell when science is much less narrowly based than now. For the present, their ideas at best suggest programmes that might engage the attention of a generation of philosophers.

Meanwhile, applied scientists and engineers have contemplated more direct approaches to the problem of expert responsibility, analogous to medical ethics. For example, in introducing a plan for establishing a professional qualification, Stanley Gill (1967), for the British Computer Society, saw it as the first, essential step in developing a code of conduct for computer personnel. They had direct access to vast quantities of information and performed unique functions in processing that information. 'All this calls for a high degree of integrity,' said Gill. Again, consider the 'Hippocratic' oath proposed by Meredith Thring (1966) for the engineer of the future: 'On becoming a Chartered Engineer, I vow to use my best endeavours to consider how my engineering skill may contribute towards the happiness of mankind. . . .' Thring thought that every engineer should take an examination at some stage in his training on the social responsibility of the engineer. The motives of such proposals are laudable, but they beg all the questions. There is still no framework in which the engineers are to judge their work, or its relevance to the 'happiness of man'; to hope for a clear-cut code, that will tell men plainly what is 'right' and what 'wrong' among the possible applications of science, is far too optimistic.

More important, at this stage, will be the admission, not only that some applications should be encouraged and others discouraged, but also that the selection is likely to be highly controversial and politically loaded. The ethics of 'pure' research, embodied in scientific institutions and behaviour, cannot cope; neither can the conventional ethics of society, embodied in existing political institutions and behaviour. Closing the ethical gap will require, not a new moral code or Hippocratic oath, but an 'applications ethics' to be embodied in new institutions and discourse, both scientific and political. It will not tell us the answers, but it may make sure that at least we ask the questions.

Part II

CULTURAL REVOLUTIONS

Mao's Cultural Revolution concerned itself, in part, with the question of whether technological progress in China depended on an elite of highly trained manpower or on workers in general. In the advanced countries, governments are increasingly obsessed with counting the heads of their scientific and technological elite. What passes for science policy is often little more than a matter of sums with money and manpower. The quiet aversion of young men from technological careers confounds the technocrats more than all the student demonstrations do; yet the threat to prosperity is illusory.

This second part of the book traces political roots and branches of some current modes of science policymaking and of scientific advice in government.

Chapter 6 deals with the relation between science and education; conditions for success in research; the supply of trained manpower; and the impending electronic revolution in education that will help answer the riddle of quality versus equality.

Chapter 7 turns to the industrial scene, in the context of European anxieties about American technological success. It looks sceptically at the fashionable and politically hazardous proposition 'You gotta be big to innovate'. But if Europeans really wish to live like Americans, they must accept the implied cultural revolution.

Chapter 8 considers the traditional sources of science advice for governments, before evaluating the current revolution attempted by the 'systems' men, both in government and in engineering.

CHAPTER 6

The Professors

THE Hungarian-born chemist, Georg von Hevesy, won a Nobel prize for showing how radioactive forms of elements could be employed as 'tracers'. On the day in 1940 when the German army occupied Copenhagen, Niels Bohr, in whose Institute he had long worked, assigned him a much more elementary chemical task: to dissolve in acid the Nobel medals of two Germans, Max von Laue and James Franck. The technical reason was that the export of gold from Germany was a serious offence. The act was nevertheless symbolic of the dissolution of German science by the Nazis.

The Axis powers paid very heavily for their persecution of Jewish and liberal-minded scientists. Pre-war American science was not particularly strong in mathematics or physics before the 1930s: astronomy and biology—at the Mount Wilson Observatory and the Rockefeller Institute, for example—were more to the taste of patrons. But Fascism, Nazism and anti-semitism drove many outstanding physicists to the USA (also to Britain) and it is indeed hard to think how the American A-bomb programme would have gone without them.

Since Hitler might nevertheless have acquired the A-bomb, and Stalin and Mao both got the H-bomb, it would be a rash assumption to think that a totalitarian regime cannot use the results of science. The scientists and engineers left in Hitler's Germany had an impressive series of developments in military technology. These included radar and infra-red devices, extremely toxic 'nerve gases', jet engines and the A4 rocket—unpopularly known as the V2, used to bombard London and subsequently the basis for both the American and Soviet missile and space programmes. But the A4 illustrates the imperfections of science harnessed to dictatorship: it was a madman's project.

To go to so much trouble and expense to lob a ton of high explosive inaccurately at a distant target made no military sense at all. The ballistic missile had to wait for the nuclear warhead to make it worth the effort.

Yet the dismal state of German research after 1945 was only in part the fault of Hitler and the war. The system of scientific universities, created in the nineteenth century, had stiffened like the necks of the professors. Those familiar with the tradition of scientific behaviour in Britain and the United States, where the young researcher is expected to outstrip his teachers, may not realize the degree of veneration accorded to professors in Germany, right up to the 1960s.

The origin was better-humoured. The illustrious French chemist Joseph Louis Gay-Lussac swept his young collaborator, Justus Liebig, into a waltz round the laboratory to celebrate their elucidation of the explosive fulminates. As a boy in his home town of Darmstadt, Liebig had been sacked by an apothecary for whom he worked, for blowing out the attic window in an experiment with fulminates. But his research on the same materials with Gay-Lussac won for him the chair of chemistry at the University of Giessen, at the age of twenty-one. The date was 1824, and it was an appointment of unusual importance because Liebig opened at Giessen the first public laboratory for the teaching of experimental chemistry. Not only can the modern pattern of higher education in science be traced to Liebig at Giessen, but the more direct effects of the establishment of schools of chemistry in Germany left untouched no aspect of world history or of everyday life.

Otto Hahn (1967), discoverer of uranium fission, related how as a young man back in Germany, after working in London with William Ramsay and in Montreal with Ernest Rutherford, he contradicted a well-known scientist who was denying the supposition that radium was a new element. The occasion was a meeting of the Bunsen Society, and one professor was heard to say to another, 'Oh, one of those anglicized Berliners'. A little later Hahn, who had discovered mesothorium, went to give some information to another eminent man. This professor referred to the material as 'semithorium' and affably contradicted Hahn when he tried to point out that the name was mesothorium. 'Remembering the bad impression I had made at the Bunsen Congress in Hamburg, I said no more.'

That was before the first World War. Equally remarkable was the demeanour of a German professor at the end of the second World War. When the Americans were rounding up the key German scientists concerned with the unsuccessful German A-bomb project, for internment in England, Carl-Friedrich von Weizsäcker complained that some of the physicists with whom he was corralled were not important enough (Irving 1967).

Hitlerism and professorism combined to thwart the German drive for the A-bomb. Of three principal routes from natural uranium to nuclear explosives, one, separation of uranium-235 by gas diffusion, was ignored—the inventor was a German Jew, Gustav Hertz. A second, production of plutonium in a graphite-moderated reactor, was not attempted because Professor Walter Bothe of Heidelberg made a grossly inaccurate measurement of the passage of neutrons through graphite and reported that the reactor would not work. The error was not detected for four years, until just before the defeat of Germany. As David Irving (1967) put it, 'Nobody had dared to suggest that a Bothe could be wrong'. That left only the heavy-water-moderated reactor as a likely route to plutonium, but entirely dependent on the supply of heavy water from electrolysis plants in Norway: it was successfully blocked by Allied commando and bomber attacks.

In 1910, half the world's chemical research was done in Justus Liebig's homeland; every ambitious science student learned German. Half a century later, the German share in world chemistry had fallen below that of France, Britain or Japan, as well, of course, of the USA; everyone was learning English. By the 1960s, the effort to reactivate German academic science had become a protracted struggle between reformers and professors. The non-university Max-Planck Society (MPG), created as the Kaiser Wilhelm Society in 1911, runs about fifty institutes (mostly small) which preserve much of the excellence of research in Germany, but which are segregated from university teaching.

When the German Research Association (DFG), the central but self-governing organization for the promotion of research in West Germany, attempted to encourage interdisciplinary research in the universities after 1962, by setting up special units, it was unable to establish satisfactory links. The DFG's list of 'priority programmes' included, besides the world fashions like computing, cardiovascular research, water resources and solar-terrestrial physics, several fields of just the kind that otherwise skipped through the coarse sieve of German academic tradition: mathematical statistics and operational research, 'cybernetics' and energy conversion. Life was not made easier for the science policymakers of the federal government in Germany by the inhibitions embedded in constitutional law against interference with the freedom of research and teaching. The Bonn government had no ministry of education and all decisions had to be agreed by eleven *Länder*. The policies open to the would-be reformers in Bonn were attempts to 'outbuild' the older universities with new ones, which may have more up-to-date ideas, to remove procedural barriers to young men embarking on research, and to encourage non-university research. But what a contrast with the United States, where the universities vied with one another to break new ground and to importune the government for the necessary funds!

To just such a contrast did the Israeli sociologist Joseph Ben-David (1968) attribute the gap in scientific attainment that had developed between the United States and continental Europe. His interpretation was that the non-governmental origins of many American universities forced the university president to convince donors about the importance of his institution and to find new markets for its services. Providing courses for veterinarians and business administrators as cheerfully as for historians or chemists, the American universities turned out the graduates the American economy wanted, and employed, in huge numbers. Paradoxical to Europeans was the way the best of the materialistic and competitive institutions of

American higher education nurtured excellent original research.

The British universities, in their aristocratic-meritocratic-juvenocratic ways, avoided the worst features of the continental system; Ben-David put them mid-way to the American system.

Nor were the British really so slow. Admittedly two decades had elapsed after Liebig opened his chemical laboratory in Giessen, but in 1848 William Thomson's (Lord Kelvin's) converted wine cellar under Old College, Glasgow, was the world's first physics laboratory for students. By the time of the critical military confrontation between Britain and Germany in 1940, it was physics, not chemistry, that counted. The twentieth-century flowering of physics in Cambridge, Manchester and London was one of the great events in the history of science, and its consequential developments in high-energy physics, solar-terrestrial physics, radio astronomy and molecular biology are with us today.

Britain is the only sizeable European country that has maintained and improved its position in world science; it has left the other Europeans, except the Swedes, well behind. In the 1950s and 1960s the scientific Nobel prizes became a rather exclusive US-British affair, with Americans taking half (a much bigger share than the Germans ever did) and the British nearly half of the remainder. In proportion to population, therefore, the British were top of this particular league table. The English never thought of themselves as philosophers and this success at the frontiers of learning was hard for them to bear, so voices were heard in crescendo saying that it was only because the nation was neglecting technology.

DOUBLE OR QUITS?

The smart boy who said to his employer, 'Look, pay me one cent this first day, two cents tomorrow, four cents the day after that while I learn the job', collected at the end of a month of twenty-three working days a pay packet of $41,943 and 4 cents. Such is the potency of any rule of doubling and redoubling over equal intervals of time. Scientific research has been growing in a similar explosive mode—the number of research workers doubles roughly every ten years, and with growing costs of research superimposed on this growth in numbers, government expenditure on fundamental research in some countries in the 1950s and early 1960s was doubling every five years.

At the time of his resignation as British minister of state for higher education and science, Vivian Bowden (1965) summed up the trend thus: 'If the rate of progress which has been maintained ever since the time of Sir Isaac Newton continued for another two hundred years every man, woman and child on Earth would be a scientist, and so would every horse, cow, mule and dog as well. It cannot go on like this much longer.'

There were angry responses. *New Scientist* itself, where Bowden's essay had been published, warned that 'a premature slackening in the development of science in Britain' would be disastrous, and said it would be plausible to look forward to five more doublings—that is to say, to some thirty times the prevailing levels—over fifty years. Desmond Bernal, the Marxist crystallographer, complained that Bowden had absorbed the 'major fallacies of the Labour Party's plan. Everything is worked out on trends and consequently there is no allowance for the positive values and particularly of the novelty of science.' But John Cockcroft pointed out that scientists knew all about the levelling off of the exponential growth of expenditure on scientific research, and that the evolution of new machinery for science policymaking was a direct consequence of this appreciation.

In its first report, the new Council for Scientific Policy, under its chairman Harrie Massey (1966), explained: 'It has been represented to us that the capacity of the national economy will not permit the growth of Research Council expenditure at present rates.' During this period, I kept asking politicians and scientists whether any directive to curb the growth of research had been made by the government. Late in 1967 the secretary of state for education and science, Patrick Gordon-Walker, assured me that no such directive had been given. He said that, for science, 'the Government's commitment is as near absolute as you can get it'. Nevertheless, the scientists serving on the Council for Scientific Policy had by then felt obliged by the prevailing financial climate in government greatly to cut back the recommended rate of growth, from 15 per cent per annum towards 9 per cent per annum by 1970.

It was really a very curious situation, in which the scientists were remarkably conscientious and gentlemanly in restraining their demands, at a time when the economic and political tide was turning against academic research. On a 'checks and balances' view of government, the CSP might have fought to maintain the growth of research. It could argue that its gentlemanly restraint was due primarily to a new awareness of limitations set by the available scientific manpower—including a swing away from the natural sciences among schoolchildren.

By the time of its second report, the CSP could also observe that a severe squeeze on research was occurring in the United States, though there was room for disagreement as to the extent this represented a real trend or merely reflected the economies for the Vietnam War. William Carey (1968) of the US Budget Bureau declared: 'The question is not whether but when the next take-off will occur in the growth curve for research and development.'

The rate of growth of the research budget is, perhaps, the one numerological aspect of research and technology that has to be taken very seriously indeed. The historical doubling and redoubling of funds for research gave expression to the talents and needs of the growing army of research workers, and opportunities for new departures in research which came thick and fast. Reduction in the growth rate will have marked effects on the character of research, not only in forcing a choice between competing programmes which in the past would both have been supported, but also by a partial stagnation, in which it will be difficult to take an important new line or to offer promotion to the brightest young scientist.

Behind all this attention to the wobbly curve of support of science lies the much bigger, unanswered political question of the long-term policy for higher education. In Britain and the United States, where the principal location for fundamental research is in the universities, the number of research workers and the funds they require are ultimately related to the number of students. While the British science advisers were agonizing about the rate of growth, I asked in vain for any indication of long-term policy for higher education in Britain to the end of the century. If Britain seriously aspired to those levels in education which are now regarded as targets in the most advanced of the United States—namely that every person should have some form of higher education—continued rapid growth of research need not look so impossible. An ambitious long-term objective might involve a tenfold increase in educational budgets by the end of the century. But this question, and the lack of an answer to it, serve only to illustrate the underlying difficulty in all science policy, that its targets lie much farther into the future than politicians are prepared to look.

FINDING THE FUNDS

If the ancient circles at Stonehenge comprise, as some astronomers now believe, a computer designed by a Megalithic Newton for predicting the eclipses of the Sun, those who think of 'big science' as a recent phenomenon are plainly in error. Observatories, botanical gardens and voyages of discovery were written into the agenda for modern science by Francis Bacon.

When, in 1805, a friend asked William Wollaston to show him his laboratory, the footman brought it in on a tray. In this century Ernest Rutherford's research group made earth-shaking discoveries with an annual budget less than contemporary laboratories spend on postage stamps. Great work can still be done at trivial expense. Yet some work always was and always will be expensive. For it, as for yachting in Edwardian style, if you have to ask what it costs you can't afford it. Life at the taxpayer's expense is not so simple and, for a research worker who needs a big instrument, it is as likely as not to be a source of grief.

At a final cost of something over £600,000 the 'Mark I' 250-foot steerable radio telescope at Jodrell Bank was, in retrospect, a remarkable bargain. It played its part in many radio astronomical adventures, including the discovery of the 'quasars' and the investigation of the 'pulsars'. It gave the unexpected bonus, in science and prestige, of tracking spacecraft; even though only a very small fraction of the telescope's observing time went to that purpose, it secured the publicity that finally saved the university from financial burden, and its progenitor from the risk of prison. Bernard Lovell (1968) has told the fraught financial history of the building of Mark I. The British government was pleased to boast of the Jodrell telescope but was unwilling to help clear the debt arising from rising costs. The Commons Public Accounts Committee (PAC) had cast blame on the consulting engineer, which it later withdrew. The debt was finally cleared by funds from private sources; not before children had been sending their pocket money to Lovell to help pay for the telescope.

This story of misjudgement and meanness was, in Lovell's view, compounded by the issue of the traditional freedom of British universities, at a time of rapidly rising public expenditure on higher education. 'It was our misfortune that the telescope was perhaps the largest and most expensive single item of research equipment that had, at that time, been undertaken by a university. Strident voices were already being raised in the PAC about these matters, and their inability to criticize university expenditure. Thus the relatively minor financial debts of the telescope came as a godsend to those voracious individuals who sought for means of gaining public control of University affairs.' A few years later, successive governments were to threaten British university autonomy, first by putting the University Grants Committee under a ministerial department (1964) and then by legislating for inspection of university accounts (1967).

In Leo Szilard's fantasy 'The Mark Gable Foundation' (1948) the narrator, revived from deep freeze in the mid-twenty-first century, is asked for advice on how to retard scientific progress. He says it would be quite easy: you set up a foundation to give grants to research workers, with committees of scientists to review the applications. 'Take the most active scientists out of the laboratory and make them members of these committees. And the very best men in the field should be appointed chairmen at salaries of $50,000 each. Also have about twenty prizes of $100,000 each for the best scientific papers of the year. This is just about all you would have to do.' The heavy irony of the story is coupled with a direct reference to the original post-war scheme for a National Science Foundation in the USA.

In a similar vein Northcote Parkinson (1962) analyzed cases such as those of Dr Tapfund who wanted to study the incidence of philately in Hong Kong and of Dr Lockstock, pioneer of research on subnormality among pop composers. He arrived at Parkinson's Law for Medical Research. 'Successful research attracts the bigger grant which makes further research impossible. In accordance with this law we mostly end as administrators.'

In spite of such satire, grant-giving machinery worked quite well. In the USA, Britain and Sweden, organization for fundamental research was not a very difficult problem in the mid-1960s, when money was plentiful. In retrospect, those were the golden years. By 1969, economies for the Vietnam War blew chill winds through many American university departments. Although the impact of the British economic crisis was less calamitous, there followed the astonishing decision of the British government, disregarding the heart-searching of its own advisers, not to support the European plan for a 300-GeV high-energy machine. There was a growing tendency of these governments to demand quick practical benefits.

Other countries attempting to catch up, and to repair the damage done by unsuitable institutions and political errors, could be comforted by the misfortunes of the leaders. These pathfinders had made no secret of their methods. The rules for successful policies for financing fundamental research, based primarily on British and American experience and promulgated by the OECD (1966), can be paraphrased as follows:

'Judgement by peers' is the crucial thing—to leave to advisory panels of specialists the decision about details of distribution of funds for research. Preferably there will be alternative agencies to which research workers can apply. The raw material in the system consists of proposals or requests from individual research workers or groups, explaining their ideas, which can then be assessed by colleagues. Is the idea original, significant, ripe for research and promising reasonable chance of success? Are the man and his institution suitable for the work, and does the proposal fall within the scope and policy of the fund-giving agency? These are the main questions. Practical benefits to be expected need not be ignored as a factor if they are apparent, but they certainly do not have to exist to justify the research. Young applicants are to be treated according to their promise, not their attainments, and relatively young scientists should serve on the panels; the membership of the panels should change repeatedly. In short, it is for the active scientists to decide what research to support. (The British 300-GeV decision grossly violated that principle.) On the other hand, governments, and particularly their more senior scientific advisers, have a responsibility for seeing that areas of inquiry of scientific or social importance do not fall neglected through interstices between the areas of interest of the agencies and their panels.

A crucial difference between American and British practice was emphasized by C. H. Waddington (1960): the frequent lack of a cushion for American university research workers, to make them independent of *ad hoc* research grants. Even though any competent research worker may obtain funds, the big prizes go to demonstrably successful ones. The result is a rat race to produce results. This is a wonderful antidote to the idleness that plagues some European academic laboratories, but who can distinguish between idleness and contemplation? In research, the most important advances may depend upon years of patient thought and experiment, with little sign of progress during that period and—most daunting of all—with no guarantee of success. Work of British physicists and chemists over two decades broached the new subject of molecular biology. While it would be dangerous to argue that such a development could not have occurred in the United States, the pressures would be against it. To this extent, Britain and Europe remain the nursery of philosophical and radical approaches to nature.

The fierce fluctuations in US congressional support for academic science, culminating in severe cuts in the budgets of many university departments in the winter of 1968–9, more than cancelled the advantage of a previously generous level of funding. Shrewdly, Richard Nixon made a token compensation to university scientists an immediate act of his new administration. American academics who envied their British colleagues' cushion of funds, not tied to specific projects, drafted a Bill to this end, promoted by George Miller, chairman of the House committee on science and astronautics.

Education is becoming the biggest industry for modern nations and its product is informed minds. Herein lies the simple justification for spending money on fundamental research. Like any other industry, education needs research to ensure that its product is as up to date as possible and not defective in any obvious way. Participation in research is itself a mode for education for more and more students. As for any other science-based industry a research expenditure of around ten per cent of the 'turnover'—total expenditure on schools and universities—seems wholly reasonable, to check that what students are told is as correct as possible. Exactly the same argument applies, of course, to research in social sciences and arts subjects. With slight modification, it explains the role of fundamental research in non-teaching laboratories of government and industry—it provides a background of education and re-education for more directly practical activities. Nevertheless, fundamental research, on this view, belongs most naturally in the university, and the segregation of that research into non-university institutes is an elementary piece of mismanagement.

CHINESE PUZZLE

Near Peking, in the late 1960s, the Chinese were building themselves a big radio telescope. It consisted of forty dish aerials, each 20–30 feet in diameter, spread out over a length of two miles. A pioneer radio astronomer from Sydney, Wilbur Christiansen, became known among his colleagues as Chris Chansen because of his visits to Peking to advise on the construction of the array. In his judgement the Chinese instrument would 'be one of the most important in the world'. Politically, though, it was a paradoxical instrument—and thereby hangs a tale.

During the Chinese Cultural Revolution of 1966–7, the disputes between supporters of Mao Tse-tung (party chairman) and of Liu Shao-chi (head of state) came to a head in the scientific as in other fields. The conflict between the quest for scientific excellence and technocratic efficiency on the one hand (Liu) and egalitarian ideals on the other (Mao) led to open fighting in research institutes as elsewhere. For the Chinese scientists, was it to be 'politics come second' or 'politics come first'?

Anyone content to dismiss the upheavals in China as a squabble between villains may miss the most graphic contemporary illustration of a world-wide dilemma, of quality versus equality. At the time of writing the West is depending on Maoist versions of the Cultural Revolution and I in turn draw on an analysis by Geoffrey Oldham (1968) of the University of Sussex, in the following picture.

In 1966, the technocratic policies of Liu, soft-pedalling politics, were dominant in the Chinese civil research community, including the Academia Sinica and its research institutes, for about five years. A Communist Party communiqué in the summer of 1966 admitted as much, by specifying in effect that scientists not actively traitorous should be left in peace to get on with their work. Yet, according to the Maoists, Liu's supporters modelled themselves slavishly on the Westerners, while Mao favoured self-reliance and would blaze a new trail for China. The Liu faction opposed allocating scarce resources to the manufacture of nuclear weapons. It had also cancelled the big radio telescope, and rationalized Mao's research programme for farm mechanization. It encouraged, the Maoists said, research divorced from the needs of the country and it wanted high salaries and scholastic titles such as 'professor' for successful scientists. From 1963 onwards, Liu's educational policy, according to the Maoists, was promoting 'elitism' and training 'intellectual aristocrats'.

The Maoists favoured national self-reliance, with everyone being regarded as a potential innovator. When they seized power in the Academy, the Maoists resumed work on the radio telescope, presumably for reasons of prestige; it was plainly divorced from the 'needs of the country'. Even so, the dispute had not been an empty one. What was better for the accomplishment of the gigantic tasks the Chinese had set themselves? Efficiency or universal struggle? Excellence in the few or aspiration in the many?

Liu's policies (as reported by the Maoists) would be widely regarded as sensible by science policymakers in the rich countries. During the early 1960s, signs appeared that China might quickly take her place among the scientifically creative nations. The exercise that in 1965 made Wang Ying-lai and his colleagues in Shanghai the first chemists to complete the synthesis of a protein (insulin) was a deliberate effort to shine in advanced research. Who knows how relevant Western judgements of Chinese accomplishment will be to the success or otherwise of that nation's progress? If the precedent occurred to the Chinese they might find an encouraging model for Maoist policies in the egalitarian and self-reliant educational and scientific policies of the US colleges of the nineteenth century.

The educational system is the principal socket whereby fundamental research is plugged into a nation's life. It also connects research with national policy and with politics. A nation relies on its educational system to provide men and women trained in various fields of learning. The choice of people for such training at high levels is a basic political act. But the pressure in every country of the world is towards more abundant education, both as a means to national strength and as an end in itself.

According to William Carey (1968) of the US Bureau of the Budget, 'The idea of higher education as a basic civil right is taking hold very firmly in the USA and is emerging as part of the national ethic'. Already two out of five American youngsters continue their education beyond eighteen. In some states the figure is higher. The aim of five out of five is made explicit by some Americans.

'THE DRAIN IN BRAINS GOES MAINLY WEST IN PLANES'

I am indebted for this sub-title to Howard Florey, late president of the Royal Society.

In 1966, immigrants constituted ten per cent of the new recruits to the American supply of engineers, and twenty-six per cent to the supply of physicians. Nearly half of the scientific immigrants were from underdeveloped countries, a loss only fractionally outweighed by the training given in the United States to scientists from the underdeveloped countries who subsequently returned home.

In Britain, where the term 'brain drain' was coined, anxiety grew during the early 1960s, even though Quintin Hogg, as minister for science, tried to minimize it and blamed it on shortcomings in the American educational system. The newspapers repeatedly carried stories of distinguished British scientists and engineers packing their bags to cross the Atlantic, and learned societies, led by the Royal Society, helped to sound the alarm. By 1967, not only had the term brain drain passed into official currency but a government working group of scientists and industrialists reported that the loss was increasing to such an extent that it constituted 'a threat to the interests of this country' (Jones, 1967). The working group rejected any idea that the outflow should be curbed by restricting movements of individuals, that some kind of payment should be made by an individual for his education, on leaving the country, or that the recruiting activities of foreign companies should be curtailed. 'The "Berlin Wall" approach offers no solution.' The right answer according to the working group, was to create more challenging opportunities, particularly in industry, for talented people. In practice, the Vietnam economies and a change in the

American immigration laws, in July 1968, promised to choke off part of the flow of young Britons.

Anthony Wedgwood Benn, minister of technology, intervened personally when the American company of Westinghouse made a special effort to recruit British experts in the technology of fast nuclear reactors, from the establishment at Dounreay, in the north of Scotland. He wrote a letter to the research workers and engineers concerned, appealing to their patriotism, and promising that their field of activity would be strongly supported by the government.

Other countries had trouble with the brain drain too. As a fraction of the output of science and engineering graduates, the loss was particularly high from Germany, the Netherlands and Switzerland to the United States. By a kind of chain reaction, there was a severe brain drain from Austria into Germany. On the other hand, that from France was particularly low.

Michel Magat (1967) attributed this relative advantage of France to a conjunction of reasons, some accidental and some deliberate. He spoke of 'healthy patriotism' and remarked: 'For most French scientists the order of moral obligations is (or seems to be): to the whole of humanity, to their country, to do science.' However, Magat saw a change in the economic situation for scientists and thought the situation in industrial research, formerly good, was growing bad. 'Actually the French industrialists never knew how to incorporate the research laboratory into the general life of the company. . . . With business slowing down, the research laboratories are reduced or directed towards short-range programmes.'

Jacob Burckhardt (1967), president of the Swiss Federal Institute of Technology, compared the brain drain of the 1960s with the 'blood drain' of the middle ages when young Swiss would go as mercenaries for the Bourbons, Hapsburgs or Prussians. The export of blood was now prohibited by law, but not the drain of brains. He reported that the Swiss engineering industries were having great difficulty in finding new staff and in some cases more than half of engineering staffs were composed of foreign nationals.

Meanwhile the American 'head-hunters' continued to use persuasive recruiting methods, and offers of excellent facilities and salaries, to draw on the talent of Europe. More spontaneously, young research workers and engineers and physicians from the poor countries of the world found opportunities in the advanced countries of both Europe and North America, and tended to linger. Hence the fear, expressed by Vivian Bowden (1965), that fields in India would go untilled in order that the Americans could put men on the Moon.

Israel lost eight hundred engineers and technicians in 1967. As a countermeasure, a government agency canvassed abroad for contracts for private companies in Israel. The idea was to keep the engineers at home, while exporting know-how. They secured a number of contracts, particularly from the less-developed countries of southern Europe and Africa. Inviting foreign-based industry to graze in the scientific pasture of the Weizmann Institute and the Technion provided a means of bringing Israelis home.

THE EXCELLENT STATES

The numerologists of science policy also began to pay attention to the geographical distribution of research effort in individual countries. For example, around 1960, about seventy per cent of all the research and development workers in France were concentrated near Paris. In the Soviet Union, a similar fraction of research workers and technologists in the institutes of the Academy of Sciences were to be found in the Moscow and Leningrad regions.

With the traditional US congressman's interest in the 'pork barrel', the fact that there were favoured and unfavoured states of the union in receipt of federal research money did not go unnoticed. An internal brain drain from Mid-Western states such as Illinois and Wisconsin to the scientific citadels of California and the North-Eastern states, was a natural consequence of a natural federal policy of allocating funds to the well-established centres of academic excellence and to the companies which already had proved their talents for technological enterprise. In the fiscal year 1965, California, with less than 10 per cent of the US population, enjoyed more than 30 per cent of the total federal obligations for research and development. New Mexico, Maryland and Massachusetts were other well-favoured states. The losers were Illinois, Michigan, Wisconsin, Indiana, Georgia and a whole string of minor states—including Oklahoma, whose Senator Fred Harris, presiding over hearings of the subcommittee on government research of the US Senate, 1966, remarked: 'It is my personal feeling that we must do nothing which will damage the existing centers of excellence in science in the country. . . . At the same time we feel that there are vast areas of the United States which are lagging in scientific development.'

By then, President Johnson had already instructed federal departments and agencies to aim to strengthen academic capabilities for science throughout the country, 'to the fullest extent compatible to their primary interests . . . and their needs for research results of high quality'.

The first big victory for the scientifically underprivileged Middle West, after the agitation started, was in securing for Weston, Illinois, the great 200-GeV machine for high-energy physics. It was 'awarded' by the Atomic Energy Commission after a competition involving two hundred locations in forty-six states. The National Academy of Sciences acted as referee as the states put in elaborate bids involving a good deal more of public relations than of science. The staff of Otto Kerner, Governor of Illinois, prepared documents weighing 165 lbs.

RELUCTANT YOUNGSTERS

Also in the mid-1960s, there was controversy about yet another kind of internal brain drain, which deflected students and young scientists away from engineering and industry and into fundamental research. Again, it was principally in Britain that the alarm was sounded, although there were echoes even in the United States.

Britain, so the story went, devoted too much effort to fundamental research and not enough of the new recruits were going into industry. Everyone in government sang the praises of technology. The young were unimpressed. The first-class honours scientist almost invariably preferred to work in university research, rather than go to industry or to teach in the schools; even among engineers and technologists the attraction of the academic world was, for the brightest, almost as strong as that of employing their newly trained talents in industry.

The fuss was misdirected. The evidence of the overseas brain drain, greatest among engineers, was that British industry was giving insufficient opportunity to the technologists it already commanded. In any case, here was another version of the numbers game, which did not take sufficient account of individual talents and excellence. The *métier* of an outstanding science student is more likely to be in fundamental research than in devising new products for export. People who are good at technology may naturally have lesser attainment in academic sciences. They are better rounded individuals: their archetype is Leonardo rather than Galileo. The evaluation of young scientists as they pass through universities is based on the standards and requirements of academic scientific research. It gives little indication at all of talents as a technologist or industrial manager. As the Council for Scientific Policy in Britain remarked (Massey, 1967a): 'We must beware of the view that the same kinds of ability and attainments are needed for the exploitation of scientific results as for original discovery.'

Young people trained in the 1960s and 1970s will be active in their professions well into the twenty-first century. Already, there is evidence—from the academic institutions of both the United States and of Britain—that the practice of technology will change drastically and quite quickly, in the direction of computer-aided design and new sociological orientations required of the engineer.

The very act of training a man as an engineer of a particular kind represents some commitment to the employment of his skill. By training large numbers of young men in the belief that their mission in life is to create new kinds of hardware, nations may be compromising social control over technological innovation for decades to come. Dennis Gabor (1967), a London physicist and inventor, warned of just this situation developing, particularly in the United States where the output of engineering graduates was soaring.

For the 'head-counters' of Whitehall, distress about the coyness of bright undergraduates was aggravated by some detective work in the schools. It revealed (horror of horrors) a growing preference for social sciences compared with the natural sciences—never mind engineering. Urgent proposals followed, for improving mathematics and science in schools and broadening senior school courses so as to postpone as long as possible so rash a choice. The kids were repeatedly making a nonsense of long-term manpower plans. To bring that into perspective, we need to look again at more important trends.

EQUALITY BY ELECTRONICS

The dilemma exposed dramatically by the Chinese Cultural Revolution is endemic in every country. In principle, universal education is the most potent egalitarian and liberalizing force of all. In practice, some schools and universities are better than others, some courses are more difficult than others, some students are more able than others. Elites are made, and the most powerful counter-egalitarian force today is the quest for excellence associated with scientific research.

Some administrators take the view that time and financial support for research in the university is just a sop thrown to the scientists to keep them contented, so that they will turn up to give their lectures. Except in the contrary sense that, if you deny research opportunities you won't get the best scientists to teach, this view could not be more mistaken. The brightest students and the most brilliant academic researcher seek one another out. They need each other and together they make the outstanding research and teaching schools.

But although research and teaching properly go hand in hand, they sometimes twist each other's wrists. At the time of the Berkeley riots in the University of California in 1965, one of the complaints of the students was that they never saw their professors—who were too busy doing research. On the European continent, on the contrary, the teaching burden on university scientists was widely regarded as too heavy. As good a rule of thumb as any is that the university scientist's time should be divided half and half between research and teaching (or, as some lament, half on research, half on teaching and half on committees).

Consider now the less capable student. If all universities and courses set the same standards, there would be an arbitrary level of attainment below which he could not qualify. A spectrum of institutions and courses is necessary if each individual is to be fruitfully educated according to his capacity. The snag is that the weaker student needs the better teacher; the better student also needs the better teacher; and the better teacher needs time for research.

Higher education for all will become one of the main objectives of leading nations in the remaining decades of the twentieth century. The means necessary for further expansion in higher education will themselves help to resolve the dilemma of quality versus equality and diminish the privileges of the elite schools, without impairing quality. The impending technical revolution in school and undergraduate education will produce a pattern more akin to the correspondence course than to the Oxford college, but one in which the influence of the best researchers and the talents of the best instructors will be available to all comers as their teaching power is multiplied by electronics. The existing styles of university will survive mainly for teaching graduates, for research and for providing the workshops for the electronic teaching material. The alternatives—growth of universities into human ant-heaps as big as cities, and continued denial of the best possible teaching to the weaker students—should be pondered by anyone eager to dismiss electronic aids to teaching as inhuman or impersonal.

For Edward Short (1969), British secretary of state for education and science, the plan for the Open University is comparable with the National Health Service, the most important social innovation of an earlier Labour government. Evolved by Jennie Lee (widow of the creator of the health service) in the face of strong political and academic opposition, the Open University will teach by television and other technological aids. It will allow any student to spend as long as he likes in graduating. In time it will offer 'post-experience' courses.

In the mid-1960s experiments in the United States on computer-aided learning for all ages reached the stage when teachers, educational authorities and the federal government began to take it seriously. Hopes began to grow in business circles for a massive American investment in the computerization of schools. Many billions of dollars might be spent on hardware (computers and terminals) and there could be a great demand for software—in not only the basic programmes needed to make the system work but, more important, teaching courses in many subjects, carefully prepared to ensure high quality. In anticipation of a boom in educational technology many affiliations of engineering and publishing firms occurred—including the pairing up of such giants as the Radio Corporation of America with Random House, and General Electric with *Time* magazine. Educational psychologists began to appear on the staffs of hardware companies, and engineers in publishing houses.

Technically, it is now possible to programme computer-aided learning systems in such a way that a child or university student can converse with the machine, learn from the machine itself how to use it, and then go through a lesson at his own pace—his progress being automatically monitored for the information of the teacher. Sound recordings, visual recordings on film and tape, books and other material can be incorporated into a common teaching system.

According to Theodore Sizer (1968), dean of education at Harvard, the education-technology combines found the going 'financially intolerable'. Yet those who anticipated the emergence of a huge new educational industry were not being naïvely optimistic. After the Moon effort, it might be the next big US federal programme. Any country which has the resources to develop computer-aided learning is also likely to be anxious about the overall quality of instruction in the schools and minor universities, and the shortage of teachers. If the new education techniques become something more substantial than a hope, during the 1970s, a cultural revolution will ensue in the rich countries of the world—very different from Mao's but having the same ultimate root.

If individuals are not forever to be regarded as fodder for industries, laboratories and offices, the idea of vocation will be eliminated from education, at least up to about first-degree level. As choice of occupation is the most important personal and political act of the ordinary individual's life, a young person's choice should not be unduly constrained by temporary needs of society. To say that is not only a matter of personal freedom. It also makes technological sense, at a time when all professions—not only engineering and technology—are in the melting pot of computerization and changing techniques. Individuals will need education and re-education throughout life. In practical terms, provided a finite fraction of students opts for each of the main lines of study and enters the principal professions, society will continue to function. After all, there are crucial areas of study (like the interactions of technology and society!) which at present are almost entirely neglected in formal education—yet society gets by.

If bright youngsters turn, say, to the underdeveloped social sciences, life in the future will be different from what it would have been if they had gone where the manpower planners wished, into engineering. Whether it will be better or worse, who can tell? But the phenomenon illustrates how education and the choices of young people promise to be important long-term factors in shaping the world and determining the quantity, quality and uses of science.

CHAPTER 7

Fiction from Rome

'WE POLITICIANS have decided that there is a technology gap —so there is one!' Thus did a junior minister from Rome chide a group of experts who were rapidly concluding that the technology gap was a misleading fiction. Taking the cue from French science policymakers who had drawn alarming conclusions from international statistical comparisons, Amintore Fanfani had invoked the gap at a NATO meeting in 1966. The Italian foreign minister went on to propose a technological version of the Marshall Plan, the post-war scheme of economic aid from the USA to Europe. Collaboration between the USA and European countries, the Italians thought, should be mounted in a range of key technologies such as computers, space operations and nuclear energy. Such a programme, they declared, 'could put Western Europe in a truly competitive position in the near future'.

Everyone dutifully started fretting about the technology gap. In Washington Lyndon Johnson and his science adviser, Donald Hornig, set up a high-powered committee to examine it and discuss it with the Europeans. In Paris the Organization for Economic Co-operation and Development (OECD) undertook a study in depth to meet a deadline for a full-dress discussion at a meeting of science ministers in the spring of 1968. A working group was set up under Jacques Spaey, chairman of the Belgian interministerial committee on science policy, and it looked into the nature and causes of transatlantic differences in specific industries.

For the Italian government, it was a boomerang. Those studies were to show that, among the leading countries of Europe, Italy certainly had a 'gap' but had no one but itself to blame. Concurrently, the OECD had been doing a separate investigation of science policy in Italy. The report of the team of experts was so adverse that the Italians refused to attend the customary 'confrontation meeting' with the team and publication of the report was blocked for many months. Eventually it was handed out to journalists in Rome by research workers who staged a sit-in protest about their conditions, in the headquarters of the National Research Council (CNR). In 1968, the Italian government at last established a ministry for scientific research, with substantial funds and powers.

Science in Italy was certainly in a morass. The schools and universities turned out very few scientists and engineers. Budgets were tiny. Professors had to supplement their salaries with other jobs, as well as doing work normally allocated to technicians and typists. Relatively little research was done by Italian industry.

Yet, in 1967, the gross national product of Italy grew by six per cent. Here was further evidence, if any were needed, that the links between research, technology, and prosperity were indirect. As the OECD study of the technology gap concluded, the US lead in certain areas of technology had no adverse effects on other countries' growth and trade performance.

Anxiety about the 'gap' was in part a matter of pride, and in part a fear for the future. By the mid-1960s, Europeans had exchanged fears of Soviet ascendancy for an acute jealousy of American success, and a feeling that they were beset by American engineering on every hand—Boeing airliners, IBM computers, Varian scientific instruments, Westinghouse and General Electric nuclear power plant, Defense Department weapons. There was a sense of dismay even about the petroleum and chemical industries, which were comparatively strong in Europe. Europeans were appalled, too, by the steady encroachment of American capital and management into the science-based industries of Europe, and looked forward with dismay to a time when American domination would become total.

A French writer Jean-Jacques Servan-Schreiber (1967) caught the mood in his book *Le Défi Américain*, remarking that while the US Army would hopefully leave Vietnam, US industry would not leave Europe. He wanted to see Europe create big industrial units that would rival the USA in management. He argued that to let the Americans carry on with their domination of science-based industry in Europe would be of short-term benefit only, not only because foreign investment limited creativity but also because the two main principles of wealth in the modern world—technical innovation and the intelligent use of resources—could be achieved only autonomously.

Ironically, when, at the beginning of 1968, Johnson moved to improve the American balance of payments of the dollar by limiting overseas spending, there were widespread lamentations in Europe. For example, John Dunning (1968), an economist at the University of Reading, commented: 'It is likely that the main effect of the Johnson proposals will be to reduce the total supply of money capital available to some of Britain's fastest growing and technologically advanced industries.' The principal investments of American companies in Europe were in oil refineries, electronics, pharmaceuticals, computers, office machinery and automobiles, bringing with them, besides cause for fear of foreign economic power threatening national sovereignty, much wanted capital and research facilities, and managerial expertise which Europe itself was lacking.

Doubts about whether European investors and managers would fill the vacuum left by the Johnson cuts illustrated the true condition of Europe, which had very little to do with technology itself. Almost as soon as consideration began of Fanfani's technology gap, there were plenty of people ready to point out that it was not a technology gap at all but a management gap—Robert McNamara was among the first to say so forcibly. Hubert Humphrey (1967), as US vice-president, echoed him in a speech at the OECD: 'If technological advance occurs more rapidly in the United States than elsewhere, the reason must be sought in educational, organizational and economic factors.' Anyone who gave attention to the problem could quickly see that the countries of Europe were certainly not lacking in engineering skills, or inventiveness—in technology proper, a branch of learning and research. A muddle in European thinking created by a confusion between technology and industry—in which Harold Wilson's Labour government in Britain was far from guiltless—served only to distract attention from the psychological and institutional factors that made all the difference between American and European industry.

In the United States, industry was obsessed with innovation and growth, and so were American inventors; their production and marketing facilities matched their enthusiasm. In Europe, not only was there a general conservatism in industry but also a lack of risk capital for new ventures. The things that were 'wrong' with European industry included a lack of science education in the board rooms and executive offices; attitudes to the role of research and innovation in Europe for several decades came behind those of the United States. Yet the European technologists and inventors were just as good as ever. If Europe wanted to make and enjoy for itself the kinds of technological products fashionable in United States industry, its businessmen would have to act like American businessmen.

That 'if' was a little breathtaking. The underlying assumption amid the alarm was that, whatever the Americans chose to make and market, Europe should imitate it; by implication, whatever way of life Americans absent-mindedly adopted, Europe should share. European nations indeed adopted American products or their European equivalents with evident satisfaction; the imitation seemed so habitual that, supposing the Americans decided something entirely bizarre, such as the desirability of living underground, Europe too would invest in the necessary tunnels and ventilation systems and solemnly move in to occupy the clay. The very conservatism and laggardly behaviour of European industry may yet turn out to have saved the continent from the full cultural consequences of imitation of the United States.

LEAGUE TABLES OF TECHNOLOGY

For better or worse, international comparisons have become the principal basis for science policymaking. There are good reasons why this should be so. Research itself is an international activity in which ideas are shared between all countries and any country can contribute ideas; there is relatively free flow of individual research workers between countries; and there is also a relatively free market in the products of technology so that any country that can afford them can buy them from abroad. There are also bad reasons, not only the 'me-tooism' already noted but also the lack of either a rational basis for thinking about investment in research and technology or the self-confidence to rely on one's own judgement.

The OECD began compiling data by which international comparisons could be made. In 1965, it published an experimental analysis of 1962 research expenditures and manpower, in Western Europe, North America and the Soviet Union (Freeman and Young, 1965). The practical difficulties were severe, both of gathering information from various countries and of correcting it for differences in educational systems and definitions of 'science'.

That report helped to arouse the French and Italian excitement about the technology gap. Among other statistics, it developed the interesting concept of the 'technological balance of payments'. The payments and receipts for technological know-how and royalties provided a crude indication of technological success. Thus the United States had receipts about eight times greater than its payments, while France and Germany, for example, paid nearly three times as much as they received—and about half their payments were to the United States alone. There were great differences from industry to industry: for example, the German chemical and petroleum products industries made payments for know-how to the United States that were less than twice their receipts; in the electrical industry, on the other hand, the receipts for know-how were one-fifteenth of the payments made to the United States. And while the West European share of new American patents was around ten per cent, the American share of European patents averaged about 17 per cent.

A couple of years later the OECD followed that report with another, based on its International Statistical Year—a study of resources, devoted to research by the OECD member countries in 1963–4 (OECD, 1967). This publication reflected more abundant and more comparable statistics, and a growing sophistication of analysis.

It gave figures for the percentage of the gross national product devoted to research and development—'as one of the most cited measures of a country's R and D effort'. By this time, for the leaders, the figures were 3·4 per cent for the USA and 2·3 per cent for the UK. The report pointed out: 'International comparisons of the GNP percentages are, however, not good yardsticks for science planning. Such a valuation can be made only in the light of the R and D aims a country sets itself, some of which are more costly to realize than others.'

Of greater significance, perhaps, was the number of scientists, engineers and technicians engaged in research and development work in relation to population. Adopting the OECD figures, one can say that while one American in every 280 is so engaged, and one Briton in 340, for Germans and Frenchmen the fraction is one in 560, for Italians one in 1700 and for Greeks one in 6700.

In the United States and France, two-thirds of the funds for research and development come from the government. Industry pays a greater share of the costs in most other countries of Europe and also Japan. Only in underdeveloped countries of Western Europe, such as Spain, Portugal and Greece, is the government share equal to that of the United States and France. As for the actual execution of the work, the norm set in the United States, Britain, Italy and Japan, for example, is that two-thirds of research and development is done in industry; industry does a greater share in Germany and less in France. In the underdeveloped countries of Western Europe, relatively little research is done in industry.

The most significant section of the 1967 OECD report is that which breaks down the research and development objectives into three broad areas: 'atomic, space and defence'; economic ends; and welfare and miscellaneous (this last including medical science and research on behalf of underdeveloped countries). It turns out that almost two-thirds of American effort goes into 'atomic, space and defence'—a much larger fraction than for the other countries compared—and only 28 per cent is economically motivated. Of course, it can be said that several aspects of nuclear and space research and development have economic motivation, and that activities in these areas and in defence have benefits in money-making industry. Nevertheless, if one subtracts that element from the United States effort and concentrates on the economically motivated research and development, the ratio of absolute expenditure on economically motivated research and development in the United States and in Western Europe is only about two to one. And as research costs in Europe are about two-thirds or less of those of the United States, the inputs into industrial research and technology give no basis for predicting a technological gap, except in the 'atomic, space and defence' fields.

The leading pioneer of international research comparisons repudiated the exaggerated conclusions drawn from his figures for 'inputs' into research. Christopher Freeman (1967), director of the science policy research unit in the University of Sussex, castigated the muddled political thinking about the technological gap and noted that the inputs were not necessarily related to the 'outputs'. There might be major differences in the efficiency of research between different countries, as between different firms. Freeman reminded his fellow numerologists: 'The study of invention and innovation cannot ignore the human imagination and the human will, however difficult measurement may seem in this field.'

Freeman also echoed others who looked sceptically on the alleged transatlantic technology gap, in saying: 'In my judgement, the really important gap, both economic and technological, is that between the poor underdeveloped countries of the world on the one hand and the richer industrialized countries on the other.'

The most important single result of the OECD's technology gap studies was to dispel the myth of the American hordes of engineers. In several European countries (Belgium, France, Germany, Netherlands) a greater proportion of young people were graduating in technology than in the USA, while a young man in Britain was *twice* as likely to take a first degree in engineering as a man of the same age in the United States. Alternatively, he was just about as likely as the American to take a doctorate in science, and the same proportion was found among young Frenchmen. Where the USA scored was in the total number of graduates, of whom only a relatively small fraction were scientists or technologists.

One OECD study looked at the origin of recent technological innovation in some representative industries. Altogether, out of a list of one hundred and forty innovations, the USA was responsible for more than eighty, Britain for twenty, Germany for twelve, and the Netherlands, Sweden and Switzerland did well in relation to their size. No discouraging gap between the USA and other OECD member countries existed in most of the industrial sectors considered:

Pharmaceuticals
Iron and steel
Non-ferrous metals
Machine tools
Scientific instruments
Man-made fibres
Bulk plastics
Electric power
Consumer electronics.

Naturally, such generalization needs to be qualified. For example, the Americans had a lead in introducing numerically controlled (automated) machine tools, in electronic measuring instruments and, among the non-ferrous metals, in tantalum and titanium. But in these sectors, comprising a large part of manufacturing industry dependent on research, the 'gap' did not really exist.

The Americans had a strong lead in innovation connected with:

Computers
Semiconductor components
Satellite communications
Specialized plastics (for defence and space purposes)
Aerospace.

This list, too, needs qualifying. Most notably, Britain had a great success in the late 1960s with specialized plastics reinforced with carbon fibres; they were developed by the Royal Aircraft Establishment and applied with immense sales rewards in Rolls Royce aero engines.

While the OECD study confirmed that it was quite common for American firms to be the first to exploit commercially research and inventions made in Europe, and very rarely the other way round, again one can cite a striking exception. In the 1950s the American physicist Bernd Matthias, of the Bell Telephone Laboratories, discovered the 'hard superconductors'. For the first time, it became practicable to make electrical machinery exploiting the fact that, at very low temperature, certain metals and alloys have absolutely no resistance to the flow of current. American efforts concentrated on making very powerful magnets for research and highly specialized devices for spacecraft. There seemed to be an 'anti-spin-off' effect of space technology here, because the first serious industrial application was in Britain, in a superconducting electric motor for a power station pump.

ROAD TO EUROTECH

In support of Britain's application to join the European Economic Community, Harold Wilson (1966) conjured up the vision of a great European Technological Community. With its lead in research, Britain was an obviously necessary member of such a community—that was the message that Wilson and his ministers lost no opportunity to deliver. Some in Britain itself wondered if they were meant to make a present of all their know-how to the French and Germans; even Europeans sympathetic to Britain's application could not quite see how else they were to benefit from British doubtlessly excellent aero engines, chemicals and fast breeder reactors—unless as compulsory customers.

There was enthusiastic talk about co-ordinated projects for computer developments, pan-European industrial mergers, a programme for designating 'centres of excellence' in different research subjects, and so on. But vagueness remained, even a year later, when Wilson developed the notion of a European Institute of Technology. His minister of technology, who had earlier spoken of a European Ministry of Technology, explained the concept of the institute as follows:

'It was not intended to be an academic institute. It was thought of as an organization which would provide an independent centre of analysis available to governments and to industries on industrial structures, problems and future development. It could undertake surveys of particular sectors of industries, of the growth of demand for their products, evaluate their potentialities and compare performances.' (Wedgwood Benn 1968b)

Others took the institute idea somewhat differently. A study group of Europeans and Americans set out a scheme for a European Institute of Science and Technology that would train leaders for the management of technological research and development. The institute would be financed by European governments and industries. The courses as outlined in the preliminary report (Giscard d'Estaing, 1968) were strongly biased towards computer applications, operational research, systems engineering and technological forecasting. The research of the institute would make it a 'think-tank' for European science and technology policy.

Before Wilson's first main speech on the subject, the research and cultural affairs committee of the European Parliament had looked to the evolution of a common scientific and technological policy between the six member states of the EEC (Oele, 1966). Science policy would serve as a link between industrial policies and the economic guide programme for the Six, which had to take account of technological trends and their encouragement. Science policymakers should strive for balance in economic and social progress and make sure that problems such as air and water pollution and the division of labour in an automated society were not neglected. While aiming to raise living standards and productivity in the EEC, the science policy of the member nations would also be bound up with the relations of the Six with the developing countries.

The European Parliament committee envisaged division of effort between the Six for projects which did not require international programming but where there was no point in duplication. But if 'community-level' research projects were appropriate, the committee did not want to urge spectacular projects like space travel; it preferred attention to basic needs of the human race. It singled out for mention housing, transport, communications and energy problems (nuclear power and coal) as possible subjects for joint projects—and added, for good measure, the suggestion of a programme of chemistry and biochemistry directed towards the development of new protein supplies for the poor countries of the world.

The reality of European technological co-operation in the late 1960s was less inspiring. The performance of Euratom, the EEC's nuclear wing, was not remarkable. The European Space Vehicle Launcher Development Organization (ELDO) was misbegotten when the British government wanted to disembarrass itself of an abandoned missile, Blue Streak. The launcher, with its British first stage, French second stage, German third stage and Italian satellite looked splendid on the pad at Woomera, Australia, but after years of difficulty, drift and dispute it was plainly of limited use. On the scientific research side, the European Space Research Organization (ESRO) invested in expensive facilities; then, just as the Americans were succeeding in launching ESRO's first single satellite, *Iris*, rising costs forced ESRO to abandon its three principal undertakings, the satellites

TD1 and *TD2* and *LAS*, the 'large astronomical satellite'. Researchers in the member countries who had been planning experiments for these satellites were disappointed, to put it mildly. Later patching saved some experiments, including *TD1* itself. Sense did not begin to prevail in European space efforts until Britain announced its withdrawal from ELDO and won support instead, late in 1968, for a unified European Space Organization to apply satellites to practical purposes. The great unqualified success of CERN, the European high-energy physics outfit in Geneva, had little to do with technology, except in the laboratories' own massive brand of scientific instruments.

European collaboration in aircraft development was more hopeful. Besides the Concorde and other bilateral projects, plans were made for the A300—an airbus jointly ventured by Hawker Siddeley, Sud-Aviation, Deutsche Airbus and Rolls Royce. There was little inclination to compete with the Americans and Russians in advanced weapons systems. The NATO nuclear planning group decided that Europe should not have an anti-ballistic missile system; Den Toom, the Netherlands defence minister, put the cost of such a system at $40,000 million.

Even granted that, in some areas of research and technology, pan-European operations were desirable, how could it be organized? Science policymaking was a primitive art. To seek to blend the heterogeneous institutions of Britain and other European nations—with their independent and state-run universities and laboratories, and their public and private industries and investments—would be a daunting prospect, even if the result had not to compete with American wizardry in technological innovation.

To be constructive, one might concentrate on creating new, largely autonomous European organizations, based on new areas of technology. Computer software, perhaps, or low-temperature engineering and superconductors, or medical engineering—these were the kinds of fields that suggested themselves. Yet these were just the technologies for which individual companies or nations would have a strong sense of possessiveness. The very greed for pay-off underlying the idea of the European

Technological Community was also the biggest obstacle to its consummation. Long before Harold Wilson, the OECD was seeking to arrange concerted programmes of applied research. Apart from speculative nuclear power experiments, it found that only the least provocative aspects of technology, like road construction, metal fatigue and noise abatement, were susceptible to co-operation on a non-commercial basis.

For all these reasons, even if Britain joins the EEC, the road to an effective Eurotech will be long and hard. And unless it intended merely to hustle Europeans faster down the Railway behind the United States, one of its more appropriate functions will be to determine what technology *not* to pursue. That possibility could not have been uppermost in the mind of Anthony Wedgwood Benn (1968a) when he spoke of 'this vital task of welding European technology into a unity which can compete with the technologies of the super-powers'.

Supernationalism built on the international ideal of science and the national ideal of foreign trade seems a questionable hybrid. But, regardless of its political merits, is it necessary?

LITTLE SWEDEN AND BIG GM

A central proposition about industry in the technological age has seized the minds of politicians, industrialists and science policymakers, especially in Europe: 'You gotta be big to innovate.' This proposition was used by government in Britain, for example, to sanction or even arrange mergers of big science-based companies which in the 1950s would have been regarded as intolerably monopolistic. It was also used in support of blanket proposals for international technological collaboration, and for British entry into the EEC.

Sweden, whose population is less than London's, has shown that the difficulties of being small can largely be overcome if one is neither too ambitious nor too timid. In the 1960s it contrived to achieve Europe's highest living standards. Its technology was impressive, too. A study by the Swedish Industrial Association in 1967 indicated that nearly half of

Sweden's recent economic progress could be attributed to technology—a contribution exceeding that of capital and labour. In electrical transmission, telephone systems, shipbuilding, tunnelling and various other fields, Sweden was very competitive. The country even developed a supersonic fighter, the Viggen, in the biggest technological project it had ever undertaken. It could not maintain more than one or two big companies in expensive fields, but the government, through its Power Board, Telecommunications Board and so on, could act as a customer for the purchase of prototype technology. SAAB was marketing a medium-sized computer, primarily in Scandinavia.

By the late 1960s, with an eye to new departures in technology, other fields ripe for development where known customers existed, were under discussion. For example, the Ministry of Education might sponsor development in educational technology, including specialized computers, and the Health Service encourage advances in medical technology. There was even talk about SAAB developing satellites, for sale to other countries; but the national prestige motive associated with satellite development made the Swedes sceptical of anyone who wanted to buy satellites at that time.

Many Swedish industrialists and technologists were very keen that Sweden, too, should join the EEC. Although the virtues of the science-based companies were visible enough, they argued for big companies for big markets, even in spite of fears that big companies could become rigid. A principal argument was the need for the marketing facilities of the big companies.

For the science-based industries, which are becoming more and more important in the advanced countries, the national market is said to be not big enough to justify the costs of major technological developments. While international collaboration may be desirable for general political reasons, the trend of this argument is decidedly weak. It smacks much more of commercial timidity on the one hand and nationalistic desires to excel at everything on the other hand, than of sober assessment of the industrial opportunities. Of course it is true that some developments in modern technology are extremely expensive. But a typically European fallacy arises from thinking of desirable technologies as a sink where money runs away, rather than as an opportunity for investment.

If a project is not worth investing in, as a national undertaking or by one of a number of firms in an industry, it will not be a good investment just because an international consortium or a big monopoly is created to do it. Conversely, if the investment is promising, what reason is there (thinking commercially) to share the benefits with other nations—unless it is, like the Concorde, unusually expensive and risky?

The argument concerning the scale of enterprise for technology goes round the problem in an orbit roughly like this:

'You gotta be big to innovate.'
'Why?'
'Because the successful US science-based companies, they're very big.'
'How do you know they're successful?'
'Because they're so big.'
'Are they big because they're successful, or successful because they're big?'
'You're missing the point.'
'What point?'
'You gotta be big to innovate.'

Consider, first, the analysis of Kenneth Galbraith (1967). For him, growth is a major aim of modern corporations. Growth naturally leads to large size, in the end.

What is not clear is whether size *per se* is indispensable to the development of a competent 'technostructure', to use Galbraith's term. Within limits, it certainly is. As Galbraith puts it, machines and sophisticated technology require heavy investment of capital and a delay between the decision to produce and the emergence of a saleable product. 'From these changes come the need and the opportunity for large business organization. It alone can deploy the requisite capital; it alone can mobilize the requisite skills.'

The income of General Motors, at more than $20 billion, exceeds the gross national product of several important countries, such as the Netherlands. How does Galbraith account for that?

'The size of General Motors,' he says, 'is in the service not of monopoly or the economies of scale but of planning. And for this planning—control of supply, control of demand, provision of capital, minimization of risk—there is no clear upper limit to the development of the size. It could be that the bigger the better. The corporate form accommodates to this need. Quite clearly it allows the firm to be very, very large.' Retained earnings provide the 'technostructure' with a source of capital wholly under its own control and capitalism, in Galbraith's view, has been fundamentally altered by the shift of power that such a source of capital brings about.

That explanation seems at first sight conclusive, and a clear indication that those who envy the Americans their successes should indeed create huge private companies, or Eurocompanies, that will busily plan how to invest their own capital. But wait—here is Richard Soderberg (1967), former dean of engineering at the Massachusetts Institute of Technology:

> 'In burgeoning industries like computers and nuclear power plants even giant corporations are running into problems of cash flow in providing the investment necessary to meet existing demand for the products of a new technology. A new and more striking example is the pending development of supersonic air transport, the SST programme. The cost of development here is so large that it is assumed that the commercial airlines and the airframe and engine manufacturers could not supply more than a fraction of the required capital.'

This is the point at which the question of size for technology can be brought down to earth. There is a spectrum of technological innovation. It starts from new science-based products that can still be pioneered and initially marketed, quite literally from a garden shed. At this lower end of the scale, either public or private enterprise could cope, but in Western countries private enterprise may be thought more appropriate. It is simply not true that 'You gotta be big to innovate': the USA itself, spawning new science-based industries around the major universities through the private enterprise of faculty members, gives evidence enough of that.

Yet some development is very costly, and then the choice of technologies open to a company depends to some extent on its size. Alan Cottrell (1968), an adviser to the British government, cited typical research and development costs as follows:

a small scientific instrument	$1 million
a new drug	$5 million
a new synthetic fibre	$20 million
a medium computer	$20 million
an aero engine	$200 million
a big subsonic jet	$500 million.

To these must be added the much greater costs of tooling up and marketing the product in sufficient quantities to cover the development costs. At the more expensive ends of the scale, the costs of some major technological undertakings are, as Soderberg indicates, beyond the capacity of private enterprise, even the American giants; the corporations must then act merely as contractors to the federal government. The real entrepreneur is the state.

Americans have gone to great lengths to conceal the role of state planning behind the façade of competition and capitalistic enterprise. The big corporations are matched to the needs of the federal government as well as to the market, so much so that the corporations, in their planning and self-financing functions, are doing what governments or state industries do in Europe. Success in technological innovation is not, in the end, dependent on giant-scale capitalist enterprise as in the United States, or even on private enterprise—which is just as well for Europe, because the investors' appetite for risk and growth is much more easily satiated in Europe than in the USA. Increasingly, the vital factor, in many areas of the world and many areas of technology, is likely to be the enterprise of governments.

I am not arguing that large-scale multinational operations are unnecessary or undesirable. They are plainly one pattern for the future. For example, a pan-European company set up to exploit one of the very new technologies, SGS (Società Generale Semiconduttori), has its headquarters at Milan, and branches in Britain, France, Germany and Sweden. Since breaking its link with Fairchild Semiconductors in the USA in 1968, it is firmly established as a European company developing and marketing integrated circuits on silicon chips. In the same field Philips, based at Eindhoven has its activities spread wide in Europe. Such commercial developments are entirely natural. What is misleading is the idea, implicit in Britain's European policy and Wilson's dire warnings about 'industrial helotry', that political aims must bow to technological compulsions—in the creation of rival superstates, with built-in pressures towards 'big' technology.

Even from the point of view of good technological management it is folly to write off the small country and the small company. In the same field of electronic components, just mentioned, one of the OECD's specialist panels on the 'technology gap' was led by Robert Galley (1969), former head of the *plan calcul* and later de Gaulle's minister for scientific research. His report emphasized the role of new companies as the electronic components business in the United States, companies like Texas Instruments, Fairchild and Motorola, which began small. The large companies in Europe were slow to see the importance, first of semiconductors and then of integrated circuits, and the Galley report stressed the need for small-scale enterprise. As for the small country, Scotland, an underprivileged part of Britain, has by regional willpower built up the largest electronic complex in Western Europe. Any country can engage in some branches of advanced technology but not all; every country, even the USA, is increasingly an obligate specialist.

A final illustration of the political paradoxes of the doctrine of large-scale technological enterprise: at the same time as the government in London was encouraging the creation of a big private monopoly for computer manufacture, the government in

Washington was preparing an anti-trust suit against IBM. The last act of the Johnson administration was to file it.

However the particular operations and modes of innovations may be broken into large or small industrial units, the technological process is nation-sized. The modern nation is an elaborate system of education and research, private capital and tax money, economic planning and communication, distribution and consumption. Even before governments began to spend fortunes on research to aid industry, they were providing industry with highly trained manpower at public expense. Increasingly, the preferences of central government help to determine the axis of industrial innovation.

Herbert Marcuse (1962) puts it more strongly. For him, modern society is totalitarian: 'The productive apparatus determines not only the socially needed occupations, skills and attitudes, but also individual needs and aspirations. It thus obliterates the opposition between private and public existence, between individual and social needs.' Marcuse therefore wants the students to smash the system.

Others may find consolation at two extremes: first, in the anarchy of individual human beings—inventors and technological entrepreneurs—who can transform (or smash) large chunks of the system from within; secondly in the growing role of government in civil innovation. The latter means a handle of power is accessible to the people for regulating and steering technology whenever they contrive to grasp it. If market factors and private enterprise were still the main determinants of innovation, it would be hard indeed to devise means of controlling the uses of science. Technology may have a built-in tendency to totalitarianism, but we cannot seriously contemplate doing without technology. Moral and political efforts need therefore to be directed to compensating the totalitarian tendency and to employing governments to encourage humane brands of technology.

CHAPTER 8

Mostly on Tap

IN JANUARY 1966, when Homi Bhabha's airliner hit Mont Blanc as it was coming into Geneva, the world's physicists mourned a brilliant colleague. Indians were stricken by the loss of the man who had given them a vision of their antique nation transformed by science. He was outstanding among the tiny cadre who rejuvenated old Mother India.

Across the harbour from crowded Bombay, at Trombay, Bhabha created one of the world's leading centres for nuclear research and development. The impact of modernity is powerful; it is an experience like Rip van Winkle's—until one notices the construction work, where lines of barefoot women carry the materials in baskets on their heads, as they did a thousand years ago.

Bhabha was pleased with his new plant at Trombay for separating plutonium from the fuel rods of his reactors, the last time I met him before his death. Its purpose was peaceful, was the line he took, but he did not want its significance overlooked, especially as the Chinese had already exploded their first nuclear bomb. India could have the bomb within eighteen months of any decision to make one. While remaining faithful to the Nehru doctrine of the peaceful atom, Bhabha was keeping the military option open technologically, as he had done politically by contesting measures proposed in the International Atomic Energy Agency for policing the nuclear activities of countries that did not already have the bomb.

During the first critical years of national independence, this Cambridge-schooled theorist, expert on cosmic rays, who had won election as a Fellow of the Royal Society of London at the early age of thirty-one, became a symbol of the new India. Abroad, he could confront the greatest scientists of the world as an equal. With charm and self-confidence he presided over the historic UN Atoms for Peace conference in Geneva, in 1955. On that occasion, physicists of East and West met for the first time since Hiroshima, and Bhabha himself dragged into public knowledge the secret work on controlled thermonuclear reactions.

At home, on his frequent visits from Bombay to New Delhi, Bhabha had personal access to Jawaharlal Nehru, his old friend and patron. He persuaded Nehru of the importance of nuclear power to India, possessing little else in the way of energy supplies, and of the need to gather the young brains of India to the task. The prime minister backed the physicist. The money, including precious foreign exchange, was provided and Bhabha swept through the bureaucracy of New Delhi like a dog through sheep.

It looked like folly. Hungry India was spending far more on nuclear research than on agricultural research; the brightest students were lured into physics rather than chemistry and biology, where the nation's more immediate needs plainly lay. There were other distinguished scientists to say so, who did their best—which was substantial—for the rest of science and technology. But with overpowering personality and enthusiasm Bhabha had his way most of the time and laid the foundations of a major nuclear power programme.

Such was the political influence of one scientist of recent times. It was exceptional. Even Frederick Lindemann, Winston Churchill's close friend and science adviser, never exerted his will in quite such a fashion; there were committees of scientists in his way, and countervailing forces in government. Lindemann's celebrated pre-war and wartime tussle with Henry Tizard, represented by Charles Snow (1961) as an heroic struggle of Folly and Wisdom, probably owed its outcome—the razing of the German cities—to more powerful political forces than either of those scientists could command; they were less like warring giants than expert witnesses, called to testify for and against the doctrine of victory by inaccurate bombing.

SCIENCE ADVISERS

Individual scientists of distinction nowadays find themselves serving as government science advisers. Alongside them are other scientists, serving on advisory committees of governments or running major scientific agencies. By the mid-1960s, a prominent American physicist could remark, 'Science has dug into the vitals of government'. In some countries, particularly in the communist countries and in brief experiments in Britain, scientists had attained ministerial rank. But the characteristic role of the scientist was advisory. The predicament he was in, and the paradoxes of his work, were roughly similar, whether he was full-time adviser to the US president or occasional adviser to the minister of tourism in Ruritania.

The notion that one man, however brilliant and broad-minded, can represent 'science' in an arm of government is more than a little absurd. Provided he has been wisely chosen, he will bring to his work the blend of inquisitiveness, imagination and awareness of ignorance that is characteristic of good science. A committee of assorted scientists backing the adviser can multiply his contacts, share the work load and stop him being merely silly. Nevertheless, every so often the adviser, face to face with the politician, has to speak for 'science' and give a clear opinion of his own. The chances of incompatibility will be relatively great, though, when the politician chooses his own adviser, there is the contrary risk that the choice pre-arranges the line of advice the politician wants to hear. The adviser is, in any case, almost certainly a 'reliable' fellow, who will neither cause needless embarrassment nor argue about purely political matters.

A scientist taken out of his laboratory and put full-time in a government office, tends to adapt to his environment: he is liable to forget his scientific attitudes and become either an administrator or a crypto-politician. In any case, he has to learn the business of government and find how he can best operate, which takes time. But if he leaves his research for more than three years or so, he will find it extremely difficult to pick up the threads again—in that sense, he pays a high professional price for serving his government. It is not surprising the scientists are reluctant to take on full-time advisory jobs and, when they do, are mistrusted by their colleagues. On the other hand, a part-time adviser may be unable to put in enough 'homework' on the government's behalf to make well-informed judgements. He has the advantage and disadvantage of being somewhat aloof from the government office.

Those who tend to remain influential in government for many years are those who manage to solve the personal equations that allow them to keep one foot in the laboratory and one in the ministry. But then they must also be able to separate their interests. Harrie Massey (1967b), simultaneously chairman of the British government's Council for Scientific Policy and head of physics at University College, London, remarked of a space research programme of his own college department: 'I have taken note of their financial plans and I am now discussing with myself whether it is sensible.'

The overwhelming problem of science advice—indeed of all political judgement about research and technology—is simple to state and hard to solve: the best-informed advice is needed for good judgement on a technical subject, yet the best-informed men are the worst advisers. The reason is expert enthusiasm. In lively branches of science it is not mere vested interest that makes the experts think their subject is the most important in the world. They genuinely believe it to be so; if they did not, the best of them would move on to something else. Accordingly, a politician wanting to know the importance of a new aircraft concept, or a possible new programme in oceanography, cannot expect dispassionate answers from an aircraft engineer or an oceanographer. Ideally, the aircraft engineer should evaluate the oceanography, and the oceanographer the aircraft. But how can they learn each other's jobs?

In practice, the pressure of specialist interests is a continuous feature of science policymaking, and attempts to compensate bring complaints. The now defunct British Advisory Council on

Scientific Policy (ACSP) was so composed that it came to be known as the Association of Chemists for the Suppression of Physics.

A leader with a forthright opinion on advisers was John Gorton (1968), who was responsible for science in Australia before he became prime minister. 'I don't know what a science policy is,' he said. 'The critics want an overall advisory committee to allocate funds, but I don't see the need for an advisory body. These committees are only a group of individuals pushing the barrow for their own disciplines.'

The science adviser will have—depending on his temperament and experience—manifest interests in particular aspects of policy. The concerns of his political chief will also affect this emphasis. Of successive science advisers of the US presidents, George Kistiakowsky (Eisenhower) was particularly concerned with the cultivation of science itself, Jerome Wiesner (Kennedy) took a strong interest in defence and foreign policy, while Donald Hornig (Johnson) reverted at first to policy for science itself. When asked in 1965 about the contribution science might make to the president's Great Society programme, his answer was, in effect, very little. He gave the impression that he had scarcely studied the possibilities. Three years later, he had found plenty of correlations, and appeared as the presidential adviser who had come to grips with domestic issues.

Hornig was not as close to Johnson as Wiesner was to Kennedy but, in 1969, President Nixon's choice of Lee DuBridge as science adviser promised better relations. Nixon had called for more military research, in contrast with Hubert Humphrey's campaign emphasis on the application of technology to domestic problems. There were hopes that DuBridge, coming from the presidency of the California Institute of Technology, would at least keep a balance.

Kistiakowsky (1965b) said: 'The scientist who becomes involved in public affairs . . . has to have a certain degree of intellectual honesty with himself and always remember there is a continuous spectrum of problems, which require very different responses from him; on one extreme are the purely scientific problems . . . on the other extreme there are purely political issues.'

But when scientist confronts non-scientist in policymaking, hair-raising scope for misunderstanding arises from their different perspectives. To a science adviser, a nuclear bomb is a weapon, a manned space flight is a prestige venture in engineering. To the politician or administrator, both are 'scientific' achievements. The scientist says they are political issues; the politician says they are scientific issues. Both are morally passing the buck.

THE ACADEMICIANS

The Royal Society of London is the oldest permanent embodiment of the scientific ethic. The new experimental philosophers of the seventeenth century evolved the ritual of experiments and meetings, organized expeditions and published the *Philosophical Transactions*, the world's first scientific journal. The original and abiding prime purpose of most scientific societies is simply a means of allowing scientists to share and evaluate the results of their work. But such societies may bring together the brightest experts in the country, and the government and the public may look to them for guidance. Then the question arises: what is the social duty of the scientific society?

It was no coincidence that the Royal Society was born in the year in which the restoration of Charles II to the throne put an end to the bitter struggle of Englishmen against Englishmen which had begun when Parliament had revolted against Charles II's father and cut off his head. Political and religious differences had shown how high theoretical ideas could lead men to mutual savagery. Experiments were 'indifferent'.

For three hundred years, the Royal Society served the community by advising the government on technical matters, but it avoided political controversy as best it could. Up to the late 1950s an 'Advertisement' appeared in each issue of the *Philosophical Transactions*: '. . . it is an established rule of the Royal Society . . . never to give their opinion, as a Body, upon any subject, either of Nature or Art, that comes before them.' This 'rule' was cautiously dropped. A retiring president, Cyril Hinshelwood (1960), reported: 'Our long and, though informal, immensely important relations with the government are in process of adaptation to the vast scale and complexity of scientific activity today.' A subsequent president, Patrick Blackett (1967), told the Queen: 'The Fellowship is widely representative of all that is best in British science and technology, and enables us to sound opinion quickly and effectively, and to voice a collective view on matters of urgent national interest.'

The Royal Society came reluctantly into direct conflict with the government over the proposal to build a military airfield on Aldabra, an atoll and biological treasure-house in the Indian Ocean. The dispute was eventually settled by a change in military policy forced by a devaluation crisis. But that controversy was still exceptional; Blackett insisted that there should be no 'reckless' entry into political controversy, and that any public expressions of views should only be made after very careful 'homework'.

If there were not a genuine dilemma about scientific integrity versus public responsibility, the Royal Society's change of policy might seem ponderously coy. In the USSR, by contrast, the equivalent Academy of Sciences ran many institutes and was closely involved in policymaking alongside the State Committee for Science and Technology. Across the Atlantic, the National Academy of Sciences in Washington was not only advising the federal government but also making pronouncements on subjects such as population growth and ocean science. The NAS, in any case, was twinned with the National Research Council, an executive science outfit. Even in its origins in 1863, during the Civil War, it was more politically oriented than the Royal Society. The first president, Alexander Bache, envisaged organized guidance to public action in scientific matters.

By the 1960s, the NAS was very active. Its committee on science and public policy (COSPUP) produced a series of elaborate studies on prospects, social connotations and practical needs in the various main branches of research. It also investigated general aspects of national science policy on behalf of Congress. The NAS reports on population, in particular, had a dramatic effect in freeing the federal government from inhibitions in dealing publicly with this subject.

Congressman Emilio Daddario said of the scientific societies (1967): 'If the societies were to come and say to us, "We would like you to find something for us to do", that would be lobbying.' By the internal revenue regulations, American scientific organizations must not attempt to influence legislation. Nevertheless, Daddario went on: 'They ought to get to the point where they make strong recommendations on issues to which their technical competence has particular relevance. It is becoming harder for us in Congress to know where the information is.'

Because of the peculiar circumstances of the rise to eminence of scientists in public affairs—with the A-bomb, radar and operational research—the engineers tended to be left out, even though much of 'science policy' was really 'technology policy'. Of the presidential science advisers, Killian and Wiesner were engineers, but the creation of a National Academy of Engineering (NAE) alongside the NAS, and talk of a similar move in Britain, testified to the sense of eclipse of the engineers. As Jay Forrester (1967) of MIT complained: 'The engineer who at one time was the educated and elite leader in matching science to society is fast becoming just another member of the industrial labour pool.'

'We are undertaking a rather grandiose national task,' said Chauncey Starr, dean of engineering of the University of California, Los Angeles, when setting up a committee on engineering and public policy (COPEP) of the National Academy of Engineering, at the end of 1966. This committee was constituted in the belief that there was a practical engineer's viewpoint on issues of public policy distinct from that of university-oriented scientists, who already had COSPUP. COPEP was to expound the social and economic consequences of the new technologies; to assess future trends in technology; to show how federal support could encourage movement in particular directions. Starr spoke of 'criteria of wisdom' as a necessary part of the education of the new engineers.

Air pollution was one of the first issues for consideration in a joint environmental studies board of the NAS and the NAE. As the engineers saw it, nuclear power production and the electric or steam automobile offered the only radical solutions. Almost unbelievably, though, there were no ecologists on the environmental studies board; they came lower than engineers in the peck-order. The engineers also took a cool look at cities. If the only strongly valued function of the city was cultural, then it would almost certainly be easier in the future to provide fast transport to the cultural centres from great distances, than to support congested cities as 'national monuments'. In 1968, an NAE committee began to look at the telecommunications of 1980; it was to advise the US president's Task Force on Communications Policy—and also the Department of Housing and Urban Development in connection with that department's study of life in cities in the future.

In Sweden, for recent historical reasons, engineers played a more important part in science policymaking than in most countries. The Royal Swedish Academy of Engineering Sciences (IVA) was created in 1919 to counteract the pull of the flourishing pure science of Sweden for the country's bright young men. The Royal Swedish Academy of Sciences was much more venerable, but the IVA became the more important source of advice for government. Under a vigorous president, Sven Brohult, IVA was very influential in Sweden in the mid-1960s. It was responsible for, among other things, international relations in science and for the science attachés at Swedish embassies abroad. In 1967, IVA created an 'investigations secretariat', concerned with technical areas of impending importance and providing a potential vehicle for serious study of the social consequences of technology. By 1968, when the government was reorganizing its arrangements for technology, some politicians began to think that the IVA had become too powerful for a non-governmental body.

The role and limits of scientific societies in public matters continued to raise doubts and dispute among their own members, too. In the winter of 1967–8, dissension occurred within the big and distinguished American Physical Society. Charles Schwartz of Berkeley had proposed an amendment to the constitution, that would set up a balloting mechanism whereby members could be polled on any issue. In arguing for it, Schwartz (1968) specifically mentioned the Vietnam War; he wanted his professional society to face up to 'an external crisis of such magnitude that we fear a general catastrophe of a political, military or cultural nature'. As one member commented facetiously: 'Before we consider Vietnam, I think we should take a ballot on the naming of element 97, Berkelium. I never really liked that name and propose we change it to Cantabrigium, which has a much nicer cant. I can think of other ridiculous possibilities.' (Kahalas 1968)

SOPHISTICATION AND SECRECY

Not only the engineers and ecologists questioned the fashionable scientists' pre-eminence in high-level advice. The American political scientist Norman Kaplan (1965) declared that 'Understanding the substance of science is *not* a sufficient basis for understanding the role of science in society'. With a growing realization that no one, not even the ablest scientists, understood that role, some scientists and social scientists began academic studies of the so-called 'science of science'. The movement gathered strength in the USA and, more slowly, elsewhere. In recognition of the need for greater sophistication in such matters, the American Association for the Advancement of Science made the political scientist Don Price of Harvard its president for 1968.

Are science advisers perhaps useless—in the sense that nothing would happen very differently if they did not exist? One British scientist with thirty years' experience of advising the government testified that he had never known a decision taken on rational grounds. In the USA the biggest civilian decision taken in this field—men to the Moon—was taken by

President Kennedy despite opposition of top advisers in Washington. And looking at the sequence of mishaps in British weapons procurement in the late 1950s and 1960s, one would have to suppose of the incumbent as defence science adviser, Solly Zuckerman, that his advice was bad—if it were not plain it was often ignored.

Zuckerman's reputation was that of a lone operator who surrounded himself with very little in the way of staff. Towards the end of his period at the Ministry of Defence, he built up the 'think-tank' at Byfleet, to help evaluate weapons systems on an inter-service basis, but his successors benefited more than he did. And when he went on to be chief science adviser to the Cabinet, with general responsibility for overseeing, with his central advisory council for science and technology, the nation's efforts in science and technology then reaching £1000 million a year, Zuckerman's staff remained minuscule. He and his council took upon themselves the most complex and difficult issues of science policymaking, without apparent resources for dealing with them in a satisfactory way. The arrangements bore no comparison with the US president's science advisory committee (PSAC) and its supporting Office of Science and Technology (OST).

Yet, amateurish though his approach seemed, Zuckerman was no fool. Brilliant wartime work on the effects of bombardment brought this authority on the sex-life of apes to the high reaches of government. He met the first test of a science adviser—that of remaining a scientist—by continuing at week-ends to direct research on reproduction in his department of anatomy at Birmingham. But his humane speeches and writings on science and society conveyed the passive sophistication of Tolstoy's Kutusov rather than the brittle cleverness of the 'think-tank'.

'Advisory bodies can only advise,' said Zuckerman (1967). 'Science policy would have much greater meaning than it has if only there were fewer unknowns in the scientific and technological process. Since the scientist is in the public arena only as the expert worker and adviser, it is his employer . . . which commands his service and which has the responsibility for action. The decision whether to accept or reject his advice is theirs and theirs only. If the scientists who now advise want more than this, then they will have to become politicians. . . .'

The nation's reliance on the good judgement of the science adviser is greatest when there is no chance for scientists outside government to check what he is saying. Among the traditional democracies, Britain is particularly prone to closed politics, with an Official Secrets Act covering what the typist had for lunch, never mind what the professor told the minister. Nowhere is it better illustrated than at the highest political level—in the Cabinet Office. Even the existence of particular Cabinet committees is a state secret. Major issues of policy, proposals for big new technological projects and programmes of international collaboration, are handled by this occult machinery.

It was into this environment that Zuckerman and his new council had to fit. At the same time, the British Parliament was setting up its select committee on science and technology, with powers to grill experts in public, and the suggestion was made (Calder, 1967a) that if the two bodies did their job well they could come into open conflict. Later, Zuckerman (1968) was to express some regret that it had not happened and probably never would. 'There is a lot of inertia—indeed, vested interest—in scientific and technological affairs, and interest is all the more likely to become vested if it is fostered behind locked doors. . . . There are many issues relating to the exploitation of advanced technology which could only benefit from a greater public understanding. Not all advances are necessarily in the public interest.'

In big issues about the uses of science, power tends to revert to the men who control the government purse. The US Bureau of the Budget takes its responsibilities seriously.

'Suppose the National Institutes of Health want to push a major programme for perfecting artificial hearts. That is the kind of thing that we would pull out for special consideration. Then we ask all the questions. Obviously, how much will it cost? But will it succeed? Such a programme raises public expectation; it cannot be stopped short, even if development costs should grow. Then if we have the machines, how many people will want them? Will there be enough surgeons to implant them? What about servicing to keep them working? And at what sort of price will artificial hearts be made available to cardiac patients—will they be only for the rich? Yes, we have to look at the ethical problems too. No one else is asking all the questions, so it's a job we have to do.'

The speaker was William Carey, the conversation was in 1967, when he was assistant director of the Bureau of the Budget. The Bureau acts as the president's general staff. It can and must take a broader view of the repercussions of federally supported science, even than the president's scientific advisory committee, with which it works closely. Any major proposal that has to be weighed by the president himself, because it transcends the normal working scope of an individual agency, in cost or in implications or both, is exposed to these searching questions of the Bureau of the Budget. In playing this regulatory part, the Bureau of the Budget has a supreme advantage over other arms of the federal government; it does not try to be 'popular' because no one loves a budget-maker anyway, and no one can do without him.

PLAYING GOD

In 1962 the Indus river flowed through Cambridge, Massachusetts. Harvard research students carried out elaborate computations of water movements in West Pakistan, as part of an intensive study of the world's largest irrigation system. At the MIT centenary celebrations in 1961, Jerome Wiesner, John Kennedy's science adviser, heard from Abdus Salam, Ayub Khan's equivalent, of huge areas of farmland in West Pakistan being thrown out of production by waterlogging and accumulation of salt in the soil. Like many irrigation systems before it, the huge network left behind by the British had begun to go badly wrong as the level of underground water rose. Wiesner despatched a mixed team of twenty American experts from the natural and social sciences, from agriculture and engineering.

The technical solution to the waterlogging and salinity problems, developed by the team, was 'vertical drainage', to lower the level of underground water with large numbers of tube wells pumping simultaneously across a large area. The calculations showed that, unless a scheme of tube wells covered about a million acres (4000 square kilometres), seepage from surrounding areas would frustrate the attempts to lower the water table.

But such a programme would be expensive. As the leader of the team, Roger Revelle remarked (1963): 'The application of the magnificent engineering works to primitive agriculture is an anachronism as striking as, say, the building of the Pennsylvania Turnpike for stagecoaches would have been.... Accordingly, our recommendations are altogether farther reaching than might have been anticipated from the original charge to the panel.'

They amounted to a plan for the agricultural development of West Pakistan, including attention to choice of crops, fertilizer manufacture, power production, pest control, training of technicians, education of farmers, better roads, textile factories, and farm machinery. Each million-acre scheme would require around $100 million initial capital. Nevertheless, the Pakistan government accepted the proposals, at least in outline. By 1968, 50,000 tube wells had been drilled by the farmers of West Pakistan.

Disregard now the particular preoccupation of Revelle's mission with salt and water, and look instead at the mixture of expertise, the modelling of the situation by computer, and the effort to take a god-like view of the interaction of physical and social activities. These were characteristic marks of a comparatively new intellectual movement among the scientists.

This movement began in Britain with the early development of radar. With military crisis approaching rapidly, there was little time for the fighting services to discover the performance and vagaries of the new technology. Most important of all, effective use of radar in air defence depended on instant coordination of the radar stations, the air force commanders and the fighter pilots. This human-cum-technical *system* was itself a necessary subject for research—'operational research' or 'OR'. Studies of this kind spread with spectacular benefits to other military functions, including civil defence, anti-submarine operations, evaluation of new weapons and the planning of seaborne invasions.

After the second World War, OR fused with other activities. In industrial operations it reinforced older ideas of work study, management study and industrial economics, with a firmer scientific base. In the military field, OR became more ambitious, assisting in the planning and selection of future weapons systems, and merging with the German idea of the *Kriegspiel*, or war game. Most important of all, the computer arrived. This machine was a tremendous asset to the OR men because many rational analyses and calculations, possible in principle, had previously been too elaborate to attempt in practice.

The term 'systems analyst' came in. It remains somewhat vague, being used in Britain for high-level computer programmers, and in the USA for some operational researchers, for some economic theorists, and for the men at the 'think-tanks' who claim to brood objectively on military, political and social affairs. I was once taken aback to find a far-out speculation off the top of my own head (Calder, 1967b) described by an expert as 'systems analysis'.

The RAND Corporation on the Pacific shoreline in a suburb of Los Angeles was the first big 'think-tank', set up by the US Air Force to analyze military operations. It made its mark in the early 1950's in a study of the deployment of the nuclear bombers of the Strategic Air Command, with particular attention to the vulnerability of the bombers on the ground. The conclusions of this study occasioned a major shift in Air Force policy for the use of overseas bases and reputedly saved the American taxpayer some billions of dollars.

This was the school from which Herman Kahn emerged with his thoughts about the unthinkable and his exquisite gradations of different kinds of nuclear warfare, so improbably subtle that only a Kahn could distinguish between them in the smoke of conflict. Of this school the British defence scientist Solly Zuckerman (1961), himself a pioneer of the OR from which systems analysis evolved, remarked sharply: 'Let us be careful not to create in a mathematical vacuum situations which are based neither on past experience of affairs, nor on any conception of the innumerable variables and factors that determine social decision either today or tomorrow.'

To 'think about the unthinkable' is not wrong; what is, in the words of an American commentator, is to attempt 'to quantify the unquantifiable, to make objective the subjective, and to involve the terminology of probabilistic reasoning where no discoverable probabilities exist' (Philip Green, 1966).

As for 'scientific' objectivity, James Schlesinger (1968), director of strategic studies at RAND, conceded to a Senate subcommittee: 'It may be inferred that a systems analysis shop attached to the office of the secretary of defence will be quite responsive to the perceptions and prejudices of the secretary and the institutional requirements of his office.'

THE ANALYSTS MOVE IN

The general ideas of systems analysis ascended to the highest levels of government in 1961. Robert McNamara, as John Kennedy's defence secretary, made them the basis for the Planning-Programming-Budgeting System (PPBS), which in 1965 was also introduced into civilian departments of the US government. 'Cost-effectiveness' appraisals of weapons systems were the best-known manifestation of McNamara's innovation, but there was more to it than that: in essence it meant continuously looking ahead to changing circumstances and needs, developing theories of the systems involved and comparing alternative programmes, before taking decisions about policies and budgets. Rather like the housewife who claims she has saved money by not buying a hat, the McNamara team would say that their procedures saved the American taxpayer $10 billion just on the decision to cancel the B-70 supersonic bomber programme.

Complex civilian systems such as cities, transport services, water supplies and education seemed to offer plenty of analogies with complex military systems. Nevertheless American civil administrators, required to adopt PPBS, found that it was much less straightforward for them than for their colleagues in the Pentagon—who, in the end, could always just issue orders and things would happen. In civilian life, people had to be persuaded; waste was frowned upon so that one could not clean the slate of existing systems; authority was fragmented so that no one, except perhaps the president and the Bureau of the Budget, had authority for planning all the interacting features of a civil system. Moreover, the necessary information for planning was often lacking. By 1969 it looked as if systems analysis in civilian government would take a very long time to evolve satisfactorily, and would do so only by compromising the autonomy of the various departments of government and of state and city administration. It was a technique that totalitarian states or 'new' poor countries might find easier to apply.

Systems analysis is a fairly powerful aid to thinking but, as it promises to seize commanding heights of government and create computable models of a widening range of human activity, politicians and public should be on their guard about three points in particular, if they do not wish to find they have absent-mindedly abdicated to the analysts and their computers.

1. Systems analysis is not yet a true science; it is a new, speculative art of government. The models of the systems analysts are simply theories, and simplified theories at that. There may be scope for a wealthy nation to make systems analysis into an experimental science—for example, by building new hospitals or new cities according to different concepts and then comparing them. But that is for the future.

2. Systems analysis may generate strong pressures to social conformity. The more comprehensive and more 'optimized' planning becomes, the greater its authority seems; the wider is the area in which citizens are told what they should or must do; the stronger the temptation to regard deviations from predicted behaviour as 'anti-social'.

3. Systems analysis can scarcely begin to cope with general human needs, priorities, values and opinions. It has to make sweeping assumptions about them. There is often lack of information about the 'social indicators' such as skills, knowledge and health of the population, their interests, and the state of their environment. Here the situation should gradually improve, as the social sciences develop and as capacious electronic stores become available to government statisticians. Meanwhile, the only generally accepted measure of value is the price tag, which is rarely appropriate for such public operations as education, health and wild-life conservation. Can a numerical measure of non-monetary value be devised to satisfy everyone? That remains to be seen. Notable work by Kenneth Arrow (1963) on *Social Choice and Individual Values*, begun at the RAND Corporation in 1948, illuminated the system of collective decision-making in a democracy but did not itself show how values were to be assigned in an analysis. Whether the quality of life can ever be rated in computable terms is, I suspect, a conundrum that will still be perplexing our grandchildren.

In the name of good and efficient government, the systems analysts will tend to form a kind of expert 'parliament' of their own. While this tendency may be constitutionally dangerous, it will provide a means of making technology serve the community better. Whatever machinery may be devised in legislatures and elsewhere for democratic control of the applications of science and technology, the executive machinery of government has a heavy responsibility. Somehow, it must investigate, quite systematically, the benefits and drawbacks of any important innovation with which the government is concerned.

The cheerful enthusiasm of a sponsoring ministry for its own pet project will be modified by attention to the views of other ministries and of their advisers and analysts. In the case of each project, be it an aircraft, a computer grid, a new railway or a new pesticide, it will be somebody's responsibility to check with every other government department and leading non-governmental bodies the answers to two questions: 'How can your activities, and the community, benefit from this development?' and 'What difficulties or dangers do you see for your activities, or the community, if we go ahead?' Complaints from a department that it was not consulted would be automatic grounds for deferring a decision on a proposal. Thus the analysts of different departments will be forced into continuous interchange and debate. This interdepartmental analytical activity could be a powerful means for avoiding the snags that every innovation is likely to entail, while at the same time encouraging ambitious thinking about new opportunities. Given a general policy of openness in government affairs, such studies could also provide parliamentarians with precisely the analyses of assumption and consequences that they need for critical judgement.

SYSTEMS ENGINEERING

Like systems analysis, 'systems engineering' is a term meaning different things to different people. It has affinities with systems analysis, but a different emphasis. It was practised before it was dignified with a special name; architects, in particular, were for long familiar with the approach. It emerged first as a specific job in the United States, in connection with military projects.

The systems engineer's task is to look at an engineering project as a whole and at the interactions of its various elements. He has to put his mind to the overall design and operation, looking for snags and limitations. Hendrik Bode (1967), of the Bell Telephone Laboratories, described the systems engineering that went into the *Telstar* satellite, one of the pioneer experiments in civil satellite communications. It illustrated one of the important aspects of systems engineering in a complex new situation—making sure that adequate science and technology are available before proceeding.

A suitable launching rocket for the satellite existed, and there were well-established microwave radio techniques on which to draw. But the experiment stood or fell on the newly available maser, the very sensitive device that the ground receivers would need to pick up the feeble signals from the satellite; other essential novelties were a high-precision aerial and tracking system, and solar cells to power the satellite. *Telstar* itself was intended as a prototype for a medium-altitude satellite communications system.

Much careful study and technical judgement went into the choice of the medium-altitude system, but within a few years a rival system triumphed—using satellites at such high altitudes that they orbit at the same speed as the Earth rotates, so that they are stationed permanently over a particular region of the Earth. The medium-altitude system may come back into its own at some time in the future when much larger amounts of communications traffic are being handled between all parts of the world. Meanwhile, as another example of the serendipity of science, we have the Bell discovery, at a *Telstar* ground station, of the sky microwave background—the best evidence so far that the universe began in a Big Bang.

Systems engineering promises to be the characteristic style of much technology in the future. For one thing, the hardware and its functions are growing more complex. A modern airliner or spacecraft is a remarkable jumble of parts and sub-systems, all of which have to work properly, severally and in concert, if the occupants are to survive. The design and programming of computer equipment for teaching children depends on many factors not dealt with in the classical engineering text-books, like child psychology and information retrieval. With automation and rising production costs, even simple manufactured items have to be designed simultaneously with the means of making them.

Nor is that all. The computer is having a direct impact on the classical forms of engineering, at least as they are taught in the most advanced schools. As an aid in the work of computation and arrangement in design, the computer serves so well that the engineer, freed from such mental routine, has the opportunity to consider many possible solutions to a given technical problem. He can, if he is so inclined, think more broadly and deeply about the purposes of his constructions and machines, and the complex relationships with human activities. The computer is itself an aid to such thinking. The new 'systems-minded' technologist prefers to be given a social goal—for example, a public transport system to satisfy certain general criteria such as the area served and the capacity—rather than just an order for a bigger bus.

The energies of a socially oriented engineering profession may be more easily switched from spectacular and technically sophisticated products like weapons systems and space vehicles, towards mundane but socially sophisticated products meeting primary human needs, such as surface transport or water supplies. The problem of building all those houses to accommodate the 3000 million additional people of the next generation, quite apart from the massive rehousing which is necessary in the poor countries and slummy enclaves, in the rich countries, should be a challenge to any bright engineer.

Such a trend is not guaranteed. John Pierce of Bell Telephone Laboratories noted with dismay the fact that 80 per cent of the support for graduate engineering education in the United States was accounted for by the Department of Defense, the National Aeronautics and Space Administration, and the Atomic Energy Commission. Herbert Hollomon (1967b), himself an electrical engineer by profession, commented: 'This situation is self-perpetuating. Almost all of the engineering professors now coming into teaching have doctorates based on this kind of research experience, and very few have had industrial experience.'

Hollomon wanted to see engineering dealing effectively with 'public goods'—schools systems, cities, suburbs, road systems, air pollution control systems, airways systems. 'The society of engineers must encompass people who are deeply wedded to the values systems of our society—men who concern themselves with whether or not engineering is worth doing at all. It is a travesty, in my view, that engineers are responsible for the design of vehicles in which many people are killed or maimed. It is a travesty that engineers are responsible for the design of industrial plants that pollute our atmosphere and our streams.'

The US programmes against poverty and unemployment, and for civil rights and better urban living, all have scientific and interdisciplinary connotations. The government's science advisers recognized that preoccupation with the interaction between technology and social change would require a new mode of co-operation between technologists and social scientists. The new Departments of Housing and Urban Development and of Transportation tried to encourage such co-operation: HUD set up an interdisciplinary Institute for Urban Development.

There are other novelties in technology that tend in the same direction as the systems engineering. An industrial offshoot from psychology and anatomy is 'ergonomics', otherwise known as 'human engineering'. Its purpose is to make sure that the machines a man works with or travels in are adapted to suit him—instead of him having to bend his body and strain his senses to suit the peculiarities of the machine. The motive is not entirely altruistic; comfort, and freedom from strain or confusion, increase a man's efficiency. Engineers are rediscovering things that seem to be known traditionally to craftsmen before the industrial revolution. When the British Ministry of Technology published 'ergonomic' dimensions and back angle for a chair for a working man, a lady in Kent measured a Queen Anne chair (more than 250 years old) which she regarded as the most comfortable in her possession, and found the dimensions and back angle corresponded exactly with the new scientific recommendations. So far from making modern man look ridiculous, that observation justifies the contemporary work and testifies to the potentiality of ergonomics for re-humanizing the environment created by industrial designers.

Need the products of industry still be so standardized? Automation does not necessarily mean the mass production of undifferentiated consumer goods which each individual must take as they are, or leave. Numerically controlled tools, for example, can make each item differ from its predecessor, and successor, at least in some simple respect. Obvious snags arise in standardization of parts and in quality control, but in future, every teapot or motor car may look different from every other. In the simplest case, production lines will merely generate randomly or whimsically varied products, within fixed limits, which the customers can then choose according to their fancy.

At a higher level of procedure the purchaser may design his own product, however roughly, and have the machine make it for him—as bespoke engineering.

Consumer research is less of a science or technology than a political movement. Quite simple tests, applied to products offered for sale to the public, allow a reasonably objective assessment of their qualities, which can then be reported to consumers. It countervails advertising and hard selling techniques to some extent, and certainly serves the choosy individual. It also influences the more alert manufacturers, by its comments. There are seeds here for something much bigger in the control of technology, if we care to cultivate them.

Meanwhile pressures of economy in industrial operations reinforce the tendency in all modern states for things to be made and organized to suit complaisant people near the peaks of the distribution curves—the median men whose needs are discernible by market research or opinion poll and whose wishes can be met smoothly and efficiently on a large scale. Technology and increasing wealth can make yesterday's majority, the pedestrians, say, become today's minority. Another defeated minority group consists of those who like to eat eggs from free-range hens or to eat wholemeal bread—tastes which modern industrialized agriculture and food-processing industries find hard to satisfy. A woman may in future be thought devious if she chooses to bear her own child. Yet all this is unnecessary. Growing wealth and technological powers should multiply, not diminish, personal choices. A moral and political test for the new engineering will be whether it aligns itself with 'the system' of god-like social planning, or with the individual, his idiosyncrasies and his psychological quirks and fancies.

Part III

PARLIAMENT OF FEARS

'THE man who wishes to preserve sanity in a dangerous world,' advised Bertrand Russell (1954), 'should summon in his own mind a parliament of fears, in which each in turn is voted absurd by all the others.'

The parliament of these next five chapters incorporates a selection of current and foreseeable opportunities and fears that arise from the uses of science and technology. They represent the unfamiliar questions with which politicians and public have to deal increasingly, and for which conventional political ideas give inadequate guidance.

Will silver iodide smoke precipitate wars as well as rain? Can one nation allow another to pollute the shared environment? (Chapter 9).

Will the oceans become a new Wild West, where claim-staking and gun law determine who owns what? Is the zone of the stationary satellites, 22,000 miles up, to be the stage of an electronic war of ideologies? (Chaper 10).

Shall we allow national computing networks to prepare an infra-structure for tyranny? Or can we exploit them in decentralizing government? (Chapter 11).

Are we to dispense with the old-fashioned way of producing children, and seek instead to improve *Homo sapiens*? Will we leave our minds alone? (Chapter 12).

And when will anything serious be done about the gulf of prosperity between nations, and within nations, created by the white man's science and technology? (Chapter 13).

CHAPTER 9

Shared Environment

AT WINDOW ROCK, Arizona, the Navajo indians were quick to put their communal accounting on a computer; they were a go-ahead tribe. And when they had been consulted about rainmaking experiments over their reservation, on the arid Colorado Plateau, they raised no objection. The Navajos were merely curious to know whether the seeding of the clouds with silver iodide smoke from an aircraft would provide male rain (thunder showers) or female rain (soft drizzle).

Despite the chronic shortage of water, rainmaking ceremonies played little part in the lives of the Navajos. For the Zuñi indians, on the same plateau, rainmaking involved their whole system of gods and ceremonials and, through them, their economic activity, social integration and political control. Not surprisingly, the Zuñis protested vigorously about local attempts at rainmaking and forced the removal of at least one ground-based generator of silver iodide smoke. Evon Vogt (1966), Harvard anthropologist who reported these reactions, contrasted them with that of a typical Texan:

'The Lord will look down and say, "Look at those poor ignorant people. I gave them the clouds, the airplanes, and the silver iodide, and they didn't have the sense to put them together."'

THE STORM-MASTERS

A hurricane is a natural engine. It uses warm water on the wind-torn ocean surface as its fuel and it converts the energy into further violent winds. American pilots flying into the eye of the storm helped to discover how the engine works; radar and weather satellites proved to be powerful aids to the research workers of Project Stormfury. The picture emerged of air warmed by the ocean and laden with water vapour spiralling in towards the eye; in a great wall of cloud, surrounding the eye like a chimney, the incoming air suddenly leapt seven miles upwards and spilled outwards across the top of the storm. As the air rose, its water vapour condensed, releasing energy equivalent, in a mature hurricane, to a megaton H-bomb going off every ten seconds. Luckily, the engine was not very efficient—only three per cent of the energy was converted into winds.

The remaining task was to discover how the spiral engine of a hurricane formed in the first place; only then, the meteorologists thought, could they seriously set about taming the hurricanes. They were not short of ideas to try. Perhaps the wall cloud could be disrupted by inducing freezing with silver iodide crystals, the favourite magic of the rainmakers. Or maybe the trick would be to cut off the hurricane's fuel supply by cooling the sea surface in its track, either by stirring up the sea to bring up colder water or by drawing an artificial veil of cloud across the sky. Some thought it would be easier to aim to guide the hurricane away from the threatened coasts, rather than to seek to disrupt it. Success in the management of tropical storms might save the USA alone some thousands of millions of dollars' worth of hurricane damage. Yet success would raise acute political problems.

Unlike the ill wind in the adage, even hurricanes bring luck to many people—with abundant rainfall on the edge of the storm and an annual quota of heat carried by decaying hurricanes from the tropics to northern latitudes. For hurricane-busting, as for other forms of weather and climate modification, international control is going to be essential if disputes are to be minimized and if the risk of accidentally starting a new Ice Age is to be averted. Meanwhile, until the meteorologists have shown convincingly that they know exactly what they are doing, the world's politicians will be well advised to discourage large-scale experiments.

During the 1960s, the meteorologists gained confidence; out of the domain of supposed cranks or charlatans, weather and climate modification passed not merely into the official research agenda but into the declarations of intent of leaders of meteorology. There were still arguments over statistics about whether silver iodide smoke really made much difference to rainfall, when it was released into promising looking clouds. But plenty of people tried it and there were few misgivings about the general proposition that puny men could deflect the mighty processes of the atmosphere by knowing where and when to tickle. Practising rainmakers claimed that twenty grammes of silver iodide could provoke the release of a million tons of water from a suitable cold cloud. Even sceptics could hardly doubt that, even if that trick didn't work, other techniques would do so, in the long run.

The World Weather Watch, a collaborative enterprise to improve weather forecasts, was agreed upon by the 129 member nations of the World Meteorological Organization in 1967. It was incidentally a step in the direction of weather and climate control. After the Americans put the first meteorological satellites into orbit (transmitting graphic pictures of the Earth's cloud cover) and as meteorologists in several countries began exploiting computers, a great leap forward in weather forecasting became possible. But it would depend upon a concerted effort by all nations to improve and intensify the conventional means of collecting basic meteorological information from surface stations, and also upon a rapid exchange of weather data of all kinds by a world-wide telecommunications network. Further novel techniques, such as the use of unmanned buoys recording the weather in mid-ocean, and balloons floating at constant levels for months high in the atmosphere, were to be brought into play. A global programme of atmospheric research was arranged to sustain the scientific momentum of the enterprise.

World meteorological centres were set up in Washington, Moscow and Melbourne and regional centres elsewhere. Costs running to hundreds of millions of dollars a year were met by the participating nations; the economic benefits to the world of better forecasts were estimated at thousands of millions of dollars.

The World Weather Watch and the associated Global Atmospheric Research Programme met a first condition for large-scale, long-lasting weather and climate control: that the planet's weather should be well observed. A second condition is that laws governing the behaviour of the atmosphere should be so fully mastered that long-term effects of human interventions can be reliably computed. Although the weather is so variable and complex that it often gives the impression of being subject to chance, successes with mathematical models of weather and the promise of extremely large computers encourage hopes that this second requirement of control will be fulfilled. One of the first customers for the huge Illiac IV Computer developed at the University of Illinois at the end of the 1960s was the National Center for Atmospheric Research in Colorado.

Thirdly, the means of intervention must be sufficient to modify the course of atmospheric activity, much as a small rudder steers a big ship. Speculative possibilities exist: the use of H-bombs partially to melt the polar ice, either in the Arctic Ocean or on the Antarctic continent, seems to be one of the easier ways of producing dramatic—perhaps too dramatic—changes in the climate of the whole Earth. Permanent regional modifications might be brought about by such means as damming the Straits of Gibraltar, to seal off the Mediterranean from the Atlantic, by modifying ocean currents or by control of mixing and evaporation of surface water. Blasting a gap in a mountain range to admit moist air to a desert beyond; or blackening a barren coastal strip to generate updraught that will force oncoming moist air from the ocean upwards and encourage clouds to form that will later rain—these are two prescriptions for arid territory. Admitting seawater into natural hollows in desert regions, like the Qattara Depression in Egypt, to form a lake, might be another. Such ideas are additional to silver iodide rainmaking, and to short-term techniques for fog dispersal and hail prevention. Tornado-busting, too, may be possible, perhaps using strips of aluminium foil to short-circuit the electric charge of the tornado funnel. Better understanding both of the large-scale behaviour of the atmosphere and of the details of cloud behaviour is likely to add new procedures to the weather-controllers' hypothetical manuals.

A fourth and most difficult condition has to be satisfied before modification is attempted on anything more than a small experimental scale. It is that everyone affected shall consent to it or, at worst, be unlikely to make trouble afterwards. The question cannot be dismissed by saying that the benefits of weather and climate control will never be worth the risks. That might be true in the temperate, well-watered, hurricane-free lands of North-West Europe, but much less plainly so in regions with pathological conditions of tempest or drought. The benefits of reliable fog-dispersal and tornado-busting would probably be incontestable. Who would forbid the use of rainmaking to relieve a deadly drought, or to extinguish a forest fire? There is no clear line of inacceptability short of massive melting or extension of the ice caps, which would flood or freeze much of the inhabited world. Several countries besides the USA are now seriously interested in the possibilities, including Australia, Japan and the USSR.

Some commentators think that you cannot make rain for one man without robbing someone else. That is true if people claim rights to particular clouds, in small-scale operations; it need not be the case in a well-planned, large-scale operation. On the other hand, the effects of weather and climate modifications are certainly not confined to meteorological events. Bringing increased rain to an arid area will radically alter circumstances for wild life adapted to the dry conditions—some species will prosper at the expense of others and so change the ecology of the area.

Anyone attempting to modify the weather in the State of Maryland is liable to three years' imprisonment. That is just one of the diverse responses of the politicians to the issues of small-scale weather modification that have arisen more frequently in the USA than elsewhere. By 1968, twenty-two other American states had enacted laws to regulate, to various degrees, interference with the weather. In addition, there was a small admixture of case law, sometimes tending to favour the experimenter but sometimes the bystander—as in a 1958 judgement of the Supreme Court of Texas, which barred hail prevention by farmers, using cloud seeding, on the grounds that by stopping precipitation they would deny moisture to neighbouring ranchers (Morris, 1966).

Despite various suggestions, and draft bills in Congress, there was, up to 1968, no US federal law dealing with weather modification which might provide a partial model for international laws. An international convention will be needed, almost certainly before 1980, to cope with widespread use of local weather-changing techniques and with proposed large-scale experiments in climate modification. The lawyers and diplomats will seek to invent a drill whereby, for example, independent meteorologists would concur on the likely consequences of any proposed action, and any affected nation could then raise objections. The convention will probably also ban the use of weather or climate modification for military purposes, when the actions affect another nation (it would be unrealistic to prohibit, say, fog dispersal from home air bases). All modification activities may have to become reportable to an international agency and there may even have to be an international monitoring network to detect silver iodide or other evidence of clandestine interference with the natural weather. Otherwise, one nation could wage a secret meteorological war against another, as described by Gordon MacDonald (1968), relying upon the natural expectation of drought to conceal a theft of rain.

If effective climate control becomes feasible, it may be an extraordinary blessing for many peoples. It will certainly have a radical effect on the tone and purposes of international relations, for better or worse. Because of the magnitude of the issue—interference with the natural environment more radical than anything attempted by men so far—it commands attention while still only a hypothesis. A much more pressing, though closely relevant activity is the unintentional poisoning of the environment.

THE EFFLUENT SOCIETY

The tanker *Torrey Canyon* carried the Liberian flag, had American owners, and an Italian captain; in March 1967 she was under charter to a British company when she struck rocks in international waters near the south-western tip of England. Much of her cargo of crude oil from Kuwait spilled into the ocean. It threatened English tourist beaches, fisheries and shellfish beds, but persistent northerly winds swept much of it on to the French coast. Scientists of both countries hastily improvised means of trying to deal with what threatened to be a regional disaster. When it was plain that the Dutch salvers could not refloat the ship, bombers attacked the ship with napalm to set fire to the oil remaining in her tanks.

The ancient laws relating to shipwreck and piracy, which the government in London disregarded in attacking a ship on the high seas, had not anticipated the advent of giant ships carrying noxious cargoes that could contaminate thousands of square miles of ocean and hundreds of miles of coastline. Commentators at the time looked gloomily at plans for building tankers much larger than *Torrey Canyon*, and at the growing number of nuclear powered submarines and surface ships which, if similarly breached, could release radioactive materials into coastal waters or harbours. An urgent meeting of the Inter-Governmental Maritime Consultative Organization (IMCO) was called, to discuss the technical and legal implications of the *Torrey Canyon* affair. Men had awoken to yet another way in which they could foul their planet, to add to an already formidable list.

A year after the disaster, biologists in Plymouth reported that the principal remedy—widespread use of detergents—had been worse than the disease, as far as marine and beach life were concerned (Smith, 1968). The *Torrey Canyon* oil was not lethal to organisms, except to the sea birds, and on rocky shores left untreated by detergents the oil was cleared naturally by limpets and other animals. But small traces of detergent were deadly to marine life. Altogether 12,000 tons of strong detergent went into the sea and on to the beaches, in the British operation. The tourist beaches were saved for the summer; if human beings had suffered more inconvenience, the damage done to the ecology of the sea and its margins might have been less quickly forgotten.

At great risk to himself and to the other species of the planet, industrial man was contaminating his environment with noisome and noxious substances of many kinds. Some, like radioactive fallout from nuclear weapons tests and pesticides used indiscriminately, had provoked intense public controversy and preventive action, but most forms of pollution mounted insidiously. From time to time governments took action against specific contaminants, but only from the mid-1960s onwards did pollution in general begin to emerge as a major political theme, notably in the United States, the countries of the Rhine, and Sweden.

Pollution of water by sewage and by industrial, agricultural and domestic wastes, and pollution of air by smokes and gases, were the foci of the new public concern. They fitted in a much broader pattern of environmental degradation and human self-poisoning, ranging from destruction of soils by bad farming practice to personal injury by cigarettes, industrial chemicals, and pharmaceuticals. Some counted noise, solid waste and effects of heat from industrial plants as forms of pollution. The havoc done by alien plants (like the Brazilian water hyacinth in African rivers), insects (like the Mediterranean fruit fly in Central America) and mammals (like the European rabbit in Australia), represented a form of biological pollution.

Nuclear energy entailed new risks, in its peaceful as well as its military uses. While a nuclear power station would not pollute the air in the manner of boilers fired with coal or oil, it generated a mess of hot, highly radioactive waste which had to be securely buried for hundreds of years, at great expense. Fuel processing for nuclear stations unavoidably contaminated water with low-level radioactivity too dilute to store.

The risk of an accident to a reactor, releasing radioactivity over a large region, can never be completely discounted, despite extreme care by the designers and operators to make it 'impossible'. In 1958, the Royal Netherlands Academy urged that international conventions should govern the safety provision and siting of reactors—pointing out that an accident to a reactor on the south-east coast of England, with a WNW wind blowing, would contaminate parts of France, Belgium, Luxemburg and Germany.

New technology can force abandonment of very old legal principles. A notable step in international legislation was the Paris Convention on nuclear liability, signed by sixteen European nations, which came into force in 1968. Its cardinal point is that of 'absolute and exclusive liability' on the part of the operator of a nuclear plant involved in an incident. In other words, anyone suffering harm as a result of a nuclear accident will not have to prove negligence on the part of the plant operator before he recovers damages. Nor can the operator try to pass on the liability to a contractor who supplied a faulty part that caused the accident. The practical motive of the men who drew up the convention in this form was fear that development of nuclear power would be held up while contractors, operators and potential victims all tried to insure themselves against the same, possibly very expensive mishaps. But the legal principle invoked to justify 'absolute liability' is probably relevant to other new technological activities: namely, that the very fact that nuclear activities are started creates a hazard that would not otherwise exist.

Now men are extending to a new domain their ability to pollute. Derelict spacecraft, rocket casings and exploded fragments have turned near space into a junk yard. The release of even small quantities of materials, such as rocket exhaust, at high altitudes could have lasting effects on the transparency of the atmosphere, and hence on the Earth's heat balance—adding to earlier speculations about the climatic effects of carbon dioxide and smoke released by burning coal and oil.

To the grief of the engineers concerned, the delicate machinery of a spacecraft destined for Mars or Venus has to be baked, to destroy terrestrial microbes that might contaminate the distant planet. The air in nuclear submarines, which stay submerged for long periods, is monitored and purified with extreme caution, lest trace gases should accumulate to injure the crew. These measures contrast strikingly with the casual maltreatment of the open environment of the planet Earth. Simon Ramo (1967) complained: 'It is a mark of our failure to control our technology that we have developed a system that can provide pure air to an astronaut on the Moon, yet we cannot do so for the citizens of our cities.'

Politicians invent laws and other means of regulating pollution. Industrialists may take action on their own account, to forestall possible restrictive legislation or claims for damages, or because of a sense of social duty. Technologists seek ways of avoiding the use of noxious materials or preventing their release. Research workers have to study the extent and nature of pollution and its effects on the environment and on living species, including man. There is thus a collective responsibility of a clarity rare in the social connections of technology. The practical question is nearly always what economic loss is deemed acceptable in the interests of limiting pollution.

Less obviously but very importantly, there is an international responsibility. The air, the oceans and a great deal of fresh water are shared. Nations releasing poisonous materials may do harm to others, cause global effects, or damage areas, like the oceans, which are nobody's property. Noxious materials are unintentionally exported and imported across frontiers; so are technological processes that release them. More constructively, technical collaboration on pollution control is appropriate because the nations have a common non-competitive interest in identifying, measuring and suppressing the various polluting agents. For all these reasons, pollution traditionally seen as a national or even as a local problem, begins to emerge as one of the prime international concerns of the immediate future. The obvious analogy is with disease, which has evoked a great deal of international exchange; pollution is a metabolic disease of industrial society.

famous for its scenery. From Lake Constance to the Netherlands it has become a gigantic open sewer. A report of the Council of Europe (1966) noted the accretion of 'germs'. In Switzerland the water contains between 30 and 100 per millilitre, increasing to 2000 by the time the river reaches Lake Constance. Many tributaries thereafter flow into the Rhine and downstream from Kembus the count is 24,000, eventually reaching more than 100,000 per millilitre.

The people of the lower Rhine, in Germany and Holland, also find that pollution upstream has added more than 30,000 tons of mineral salts to the daily flow of the Rhine. In Alsace alone, 15,000 tons of residual potassium salts are discharged into the Rhine every day, and the Ruhr mines discharge further salts. By the time the water reaches the Netherlands, it is virtually useless. As recently as the end of the nineteenth century, more than 100,000 salmon were being taken regularly from the Rhine in the Netherlands each year. The Council of Europe recorded: 'A few survivors linger today, but they are quite inedible because of their taste.' That was before the incident of June 1969, in which the release of a small quantity of a pesticide into the Rhine killed untold millions of fish.

As a measure of personal involvement with water, it constitutes 60 per cent of the weight of the human body. It is not only drinking water that is put at risk by pollution: heavily polluted water may be unsuitable for irrigation, for farm animals, and for industry. Clean fresh water has aesthetic and leisure values. Rushing rivers and calm lakes are among the greatest of tourist attractions; besides boating, swimming and water-skiing on inland waters, angling is a leisure occupation for many millions of people in industrialized countries.

Yet into the water goes solid matter, from mineral extraction, canneries, paper-pulp mills and other industries, to make the river murky and clog the gills of fish. Chemical discharges include chlorine, cyanide, phenols, fertilizers, weedkillers, fungicides, insecticides and rat poisons. The putrefying organic material includes domestic sewage. Now we add radioactive pollution.

Eutrophication, as the biologist calls it, means an excessive fertilization of water with nutrients, mainly phosphorus and nitrogen, that can lead to an overgrowth of aquatic life. It can quickly change pure water into an odious weed-choked, decaying system, unsuited for angling or for drinking. Lake Michigan has been gravely disturbed in this way, and new species have been accidentally introduced by the opening of a canal. Even if all pollution were stopped at once, restoration of Lake Michigan would take a century.

Purification can be a costly and complicated business. In West Germany, for example, investment in sewage treatment plant from the mid-1960s to the mid-1970s is estimated at more than $2 billion. Control can result in the accumulation of vast amounts of sewage sludge; why it is not used more widely in Europe and North America as an agricultural fertilizer is hard to tell. Pollution can be a valuable source of other materials: for example, Vitamin B_{12} is extracted from domestic sewage in Chicago, while coal dust, filtered from the River Emcher before it enters the Rhine, provides fuel for the local power stations. Ordinary treatment of water can kill and remove some of the noxious contents of polluted water, but it cannot deal with many of the small molecules introduced by industrial and domestic and agricultural practice, nor dispose reliably of dangerous viruses.

In the case of the Rhine, the international aspects of pollution are plain enough. But consider, next, that acid rain falling on Sweden, allegedly as a result of air pollution blowing over the Ruhr. It damages Swedish forests, and this same source of pollution was blamed for some of the mercury that accumulated in Sweden's famous lakes. Another source of mercury—in chemical dressing for seeds—was outlawed in Sweden, but the mercury content of the lakes had risen so high that, in 1968, the government had to advise Swedes not to eat more than one meal a week of freshwater fish.

Some plants are much more sensitive than animals or human beings to noxious gases in the air; that appeared clearly in the farms and forests of heavily polluted Bohemia. Plants can therefore provide an early warning of the impending injuries to ourselves as a result of the pollution in the atmosphere. In a sorry illustration of the lengths men must go to, to cope with their own activities, the US Department of Agriculture's forest service has been breeding a range of white pines, each sensitive to different polluting agents; they can then be planted strategically to monitor pollution.

The high sulphur content of some fuel oil used in power stations has made it a special target for pollution controllers. In New York, a notoriously polluted city, the power producers agreed with the air-pollution commissioner to use no fuel oil containing more than one per cent of sulphur; this single change, made in 1968, was expected to cut the amount of sulphur dioxide in the air by a quarter. The sulphur content of oil varies widely from oil field to oil field; crude oil from Middle East fields has in general a particularly high sulphur content. A general aversion from sulphur-rich oils would therefore have a big effect on trade patterns and therefore on geopolitics. On the other hand, discovery of cheap means of removing the sulphur from the fuel or the smoke would nullify such a trend.

Petroleum fuels imperfectly burnt in road vehicles have for long been criticized as sources of air pollution, especially in Los Angeles and other American cities. Beginning with the 1968 car models, the US government embarked on a progressive series of regulations to reduce noxious components in vehicle exhaust. But the Public Health Service expected to be able only to keep the total pollution pegged at the 1967 level because, although each vehicle might be less offensive, the number of vehicles was expected to increase. Particular attention has been paid to carbon monoxide, nitrogen oxides and cancer-inducing hydrocarbons in road vehicle exhaust. But high concentrations of lead are detectable in the neighbourhood of busy roads and in the blood of American city dwellers—it comes from the anti-knock agents added to gasoline.

European and Japanese car manufacturers exporting to the United States have to conform with that nation's regulations for control of exhaust pollution by special technical devices. This illustrates another international aspect of pollution, and the desirability of agreed standards in trading countries.

Common sense is not a good guide to the hazards of pollution. It suggests that pollutants such as radioactive fall-out and DDT dispersed to the air will tend to spread and become more dilute. What could seem safer than to put radioactive waste in the sea, where it would be greatly diluted by the immense volume of ocean waters? But that supposition is dangerously misleading, because it overlooks the nature of interactions between different species and between the species and their environment.

If a caribou eats lichen containing radioactive caesium-137, coming from a nuclear explosion and deposited by the rain, the concentration of radioactivity from this source in the caribou is considerably larger than in the lichen; if an eskimo then eats the caribou he will build up an even greater concentration. Such an effect occurs, not with all elements, but particularly with those which, like caesium-137, are distributed in the flesh. Similarly, in water contaminated with DDT insecticide, a minnow feeding on plankton which has relatively little DDT in it, will concentrate the chemical to such an extent that it has, by weight, twenty-five times as much as in the plankton. A gull feeding on the fish in the same lake may concentrate the DDT by another factor of 75, so that it finishes up with 1000 times as much DDT as in the same weight of plankton. Discussing these results, George Woodwell (1967) commented: 'What has been learned about the dangers in polluting ecological cycles is ample proof that there is no longer safety in the vastness of the Earth.'

The most significant political features of pollution control are, first, central regulation of activities hitherto considered the exclusive concern of local or provincial authorities; secondly, the way in which 'nationalization' of the problem is liable to be quickly followed by 'internationalization'. The operation of a German mine may be a matter of direct concern to the Dutch; the fuel burnt in a German power station may be a matter for diplomatic interventions from Scandinavia.

'Nationalization' became plainly evident in the arrangements for control of water pollution in Europe between 1950 and 1964. Belgium, Britain, Switzerland and France each adopted comprehensive legislation on control of water pollution. The

French law, one of the most up-to-date, created a national water committee, including hydrologists, biologists and other scientists as well as representatives of national government and of people with commercial interests in water. The law set a two-year deadline for a survey of surface water pollution in France. It also gave the state new power over non-navigable waterways; in special water-planning areas, public authorities could regulate the use of all waters and extinguish rights. The first comprehensive water pollution legislation in the United States was enacted in 1956, and strengthened in 1961. The legislation made clear that there was to be a strong federal role in water pollution control and assigned primary responsibility to the Department of Health, Education and Welfare.

Edmund Muskie tried repeatedly to create a US Senate committee on technology and the human environment, to keep an eye open for new dangers. As politicians awoke to the manifold threats to the environment, Richard Nixon set up a task force, which urged him to appoint a presidential assistant for environmental affairs.

At a conference in Dubrovnik in 1956, the International Law Association agreed on principles of international law which affirmed that a state was legally responsible for substantial damage done to another state by preventable pollution of water. An international commission of the five countries concerned in the protection of the Rhine was created in May 1965 to study the possibilities of a treaty. A European Water Charter was proclaimed by the Council of Europe (1968); it declared: 'When used water is returned to a common source it must not impair the future uses, both public and private, to which the common source will be put.' But its international provisions were weak, saying only that disputes should be settled by 'mutual agreement between the States concerned'. International machinery for pollution control has lagged behind recognition of problems; before long pollution control will be added to the growing list of technological activities in which international authority is exercised.

CHAPTER 10

New Frontiers

BLACK MOSLEMS should be advised that Mecca is receding from Africa at a speed of about two centimetres a year. According to current geophysical theory, the continents drift like rafts across the face of the globe, and Africa is swinging away from Arabia, slowly widening the Red Sea. A consequent rift in the floor of this incipient ocean makes the mouths of well-informed investors water like the very sea itself.

During the 1960s the prospects of fish-farming in ocean waters and of mining under the seabed provoked growing speculation. Of most obvious immediate interest were the nodules of manganese ore littering the ocean floor, which could be recovered with suitable grabs, but there was doubt about the existence of other minerals of significance. In 1966, the American research ship *Chain* made an intensive study of a remarkable 'pool' of hot, salty water at the bottom of the Red Sea, near Mecca—a phenomenon first noticed by oceanographers three years earlier. It was 7000 feet below the sea surface, in the rift valley, a region of major geological activity. Rock samples from the bottom of the hot pool revealed a treasure-house of minerals. The top 30 feet showed gold, silver, zinc and copper, later valued by the US Geological Survey at $1500 million dollars. There was also much manganese and iron; furthermore, the minerals were suspected to reach 300 feet down. Other hot pools exist in the Red Sea, and other ocean rift valleys were known elsewhere in the world. Suddenly, mining of the floor of the deep ocean had become an interesting possibility. A company called Red Sea Enterprises set itself up in Liechtenstein and laid claim to a portion of the seabed; so did the adjacent states, Saudi Arabia and the Sudan.

THE SEASQUATTERS

'Our country would suffer irreparable harm if jurisdiction to the sea bed were transferred to any organization.' Alton Lennon, North Carolina congressman and chairman of the House subcommittee on oceanography, thus testified uncompromisingly on behalf of political, industrial and military men who regarded with suspicion any action that might limit the freedom to exploit the resources of the ocean as Americans saw fit.

No aspect of civil technology seemed more likely to separate the patriots from the internationalists than the approach to the mineral resources of the deep oceans. No moral or political issue of greater consequence for the immediate future of the human species was visible in the late 1960s, than this question of how the wealth from newly accessible resources of the oceans would be distributed among men. Those who had failed to appreciate the recent advances in ocean technology (and they included most of the world's political leaders) were ignorant of the impending battle. The first skirmishes were occurring in US congressional hearings and in an inconclusive debate in the United Nations' General Assembly, on a proposal by Arvid Pardo of Malta for laws governing the floor of the ocean.

As if in defiance of the ocean resources lobby, Lyndon Johnson declared in 1967 that 'we must ensure that the deep seas and ocean bottoms are, and remain, the legacy of all human beings'. The Marine Sciences Council, of which his vice-president, Hubert Humphrey, was chairman, had assigned top priority to promoting international collaboration. Arthur Goldberg, as US ambassador at the UN, gave a friendly, though noncommittal, welcome to the Maltese proposal.

But in the US House of Representatives hearing on jurisdiction over ocean resources, the opposition to any such move to internationalize the seabed was overwhelming. The National Oceanography Association (which represented private companies engaged in ocean technology), the US Chamber of Commerce, the American Legion and Ronald Reagan, governor of

California, all registered protests against the Maltese proposal. The subcommittee on international organizations and movements (chairman, Dante Fascell, 1967) recommended in an interim report:

'. . . 2. That the US Government actively discourage any action to reach a decision at this time with respect to the vesting of title to the seabed, the ocean floor, or ocean resources, in any existing or new international organization; and

'3. That the US Government, while continuing to encourage and support constructive international co-operation in the exploration of the oceans, proceed in this field with the greatest caution so as not to limit or prejudice our national interests in the exploration, use and economic exploitation of ocean resources. The United States should urge further study of all the issues and problems relating to this entire subject.'

Two out of ten members of the subcommittee, Donald Fraser and Benjamin Rosenthal, dissented. They concluded:

'. . . we believe the United Nations should be encouraged to consider the possibility of financing its own activities, particularly those involving assistance to the less wealthy countries, from a planned development of the ocean's resources. Whether this possibility could one day solve the United Nations' financial problems is not yet known but, once again, the world assembly is the proper forum for exploring the benefits which could flow from such an international approach.'

By that time, international co-operation in oceanographic research had developed well. An International Indian Ocean Expedition had carried out thorough work in what was previously the least known of the world's oceans and many discoveries were made—including the early Red Sea findings. American, Soviet, British, Japanese and Australian research ships had taken part. It was followed by joint investigations of the Kuroshio current and the tropical Atlantic, and by plans for international projects in the Mediterranean, Caribbean, North Atlantic, South Pacific and Antarctic waters. The research workers had their Special Committee on Oceanic Research

(SCOR) of the International Council of Scientific Unions. In 1960, UNESCO set up an Intergovernmental Oceanographic Commission, which fifty-eight nations had joined by 1967.

Co-operation in research is one thing; agreement on the use of commercially valuable resources, quite another. The near-extermination of whales in the mid-1960s by the excessive hunting of the Norwegian, Russian and Japanese fleets, despite the urgent warnings of marine biologists and the existence of international machinery for setting limits to catches, gave a sombre warning of likely commercial attitudes to ocean resources that were there for the grabbing until exhausted. It also, from another point of view, served as a reminder of the vulnerability of marine life. A massive human invasion of the undersea, in search of minerals and food, could be as disastrous as activities that had blighted land areas. It may be that big increases in food supplies from the seas, whether by more efficient gathering and hunting or by 'farming' techniques, should not be attempted until we know very much more about the ecology of life in the sea.

The seabed question shows how quickly international law can be outdated by advances in technology. As recently as 1964, a new convention came into force which gave states sovereignty over minerals in the continental shelf adjoining their coasts. It allowed, for example, the carve-up of the North Sea and the subsequent discovery and exploitation of deposits of natural gas under the seabed. But that convention did not fix a clear outer limit to the rights—certainly they went as far as the 200 metre depth contour, but then as far beyond again as exploitation of the natural resources of the seabed was possible in practice. The lawyers had in mind the continental shelf, believing that the bed of the deep oceans would be inaccessible. They did not anticipate the technological advances that altered the prospects in the course of a few years. When the continental shelf convention was drafted no one thought much of the mineralogical charms of the deep oceans. Apart from those nodules of manganese ore, most of the floor of the deep ocean seemed geologically dull, unmodified by the natural processes that make useful ores. By the time that picture was transformed, by the discoveries in the Red Sea, submersibles and other deep sea technologies had proved their worth, by manned descent of the bathyscaphe *Trieste* to the deepest deep of the oceans, and in the disasters involving the submarines *Thresher* and *Scorpion* and the lost H-bomb off the Spanish coast.

While American industry sought new tools for working under the sea, the US Marine Sciences Council was co-ordinating national programmes amounting to more than $500 million. Ocean science and technology were among the very few federal activities not curbed by the Vietnam War. The military share of the effort was large but diminishing. A report by the National Academy of Sciences had estimated that for civil research sixty new oceanographic ships would be needed in the period 1966–77, together with big shore facilities and research laboratories. No other nation rivalled the American effort, although the Russians and the Japanese had substantial programmes and the French a growing and well-publicized one. The British toyed with the idea of a national ocean programme, a sort of mini-version of the US space programme but dedicated to ocean technology; meanwhile, the newly constituted National Environment Research Council in London made oceanography research a major concern.

Very different possibilities, accessible to human choice, open up before the politicians and scientists, not as some hypothetical topic for the future but one probably requiring decisions in the early 1970s. Consider two pictures.

1. The resources of the oceans belong to those who can take them, and submarine colonialism prevails. A new frontier is open to human enterprise, like the American Wild West in the nineteenth century, but now the miners and ranchers are backed by advanced research and technology and the resources of modern industrial states. Americans, Russians, Japanese and Europeans vie with each other to stake claims, in a bigger free-for-all than was witnessed even at the peak of the European empires during the grab of Africa. Undersea navies defend the squatters' installations and at the same time build up strategic

systems of their own. The merchant fleets, research fleets, fishing fleets, mineral fleets and naval fleets of the nations intermingle in mutual provocation. Fortunes are made from precious metals and trigger fingers itch.

2. The resources of the oceans belong to the United Nations Organization, which leases them to individual nations or companies in return for substantial royalties on the wealth extracted. Nations that are not members of the UN (like the Germanies and China at present) have equal standing in this respect with those that are. To encourage exploitation and enterprise, rights are normally assigned to discoverers, provided that no one nation has more than one third of the world total; limits to production may be set to maintain price levels. Theft of resources and misuses of the ocean floor for military purposes (banned by treaty) are prevented by a UN fleet provided jointly by all nations engaged in ocean exploitation. The royalties—running at several billion dollars a year in the 1980s—go primarily in aid of developing nations, but they also give the UN organization itself a financial stability that it has not previously enjoyed.

These are the extreme possibilities, but both are openly discussed and neither is far-fetched. A number of intermediate possibilities exist, including the drawing of complex lines on the map to share the oceans between contiguous states, as the North Sea was divided. But there is a pivotal question: Are the deep oceans common property to all men, or the property of no man and therefore accessible to anyone who seizes them? The answer will set the tone for international relations for the next hundred years.

Internationalization of the seabed, if that is the choice, will not be easy to secure. Apart from opposition from commercial interests, there will be plenty of room for argument about the powers and competence of the supranational authority. Because of existing naval commitments in the oceans, rules against some military activity may involve actual disarmament measures—with all the political difficulties that those entail. The parallel with disarmament goes further: agreement about the oceans will depend on the simultaneous consent of the major powers, yet it can be frustrated, as can disarmament proposals, by any one of the major powers yielding to internal political forces opposed to the agreement. Failure is on the whole more likely than success.

For that reason, simple-minded optimism about international amity in the oceans is certainly misplaced. It may be secured only by unusually strong and rapid diplomatic action. It must probably take effect before any nation unilaterally claims an oceanic region (as Harry Truman appropriated the continental shelf in 1945) or commits itself to the construction of expensive military or commercial facilities on the ocean bed.

THE SPACE RACERS

Luna 10 broadcast the *Internationale* from orbit round the Moon; the *Apollo 8* crew read *Genesis* off fireproof paper. The political pudding left behind in the human break-out into space was a curious blend of good and bad ingredients. They were admixed with great issues that, more than a decade after *Sputnik 1*, political leaders and the general public had scarcely begun to notice. Here are some contrasts:

1. Rivalry for prestige in civilian space activities continued between the super-powers, scarcely modified by years of talk about co-operation. Nevertheless, the extent of the international collaboration in space research was striking.

2. Manned space flight, in orbit and to the Moon, was a project of little immediate value to scientific research and even less to everyday life. Yet, concurrent developments of satellites observing the Earth for weather and other phenomena, and of satellites relaying messages and television pictures, gave immense opportunities relevant to life on Earth.

3. A major arms race and intelligence war in space involved spy satellites, military communications satellites and navigational satellites, with anti-satellite missiles standing ready. Even so, one of the few positive measures of arms control was the outer-space treaty signed by the USA, the USSR and sixty other nations, in 1967. It prohibited the deployment in space of

nuclear weapons or any other weapons of mass destruction, established that no state could claim sovereignty over the Moon or anything else in space, and clarified some elementary rights and liabilities of spacefaring nations. It also provided for international consultation before any experiments took place that might be harmful to the exploration of space—like the notorious *Starfish* high-altitude nuclear explosion in 1962 which radically altered the newly discovered radiation belt around the Earth.

The social and political consequences of spaceflight are so abundant that they are worthy subjects for other books. Attention is here confined to two themes only: possible US-Soviet co-operation in space and problems looming up with communications satellites.

Early one morning in April 1967, the Soviet spaceship *Soyuz 1* fouled its parachute during landing and crashed, killing its pilot Vladimir Komarov. This first death of a spaceman in flight came three months after a fire during tests of an *Apollo* spacecraft at Cape Kennedy, which resulted in the death of three American astronauts. James Webb, head of the US civilian space programme, responded to the news of Komarov's death by asking whether Russian and American lives might have been saved if there had been full co-operation. 'I very much hope that the dramatic events which have already occurred in 1967 will be looked at against the background of the many statements made by leaders of both nations to the effect that co-operation is something both nations should seek.' (Webb, 1967)

Irrationally, then, the inevitable death of a spaceman provoked introspection about the temper and tempo of manned spaceflight, and eroded a little more the assumptions of intense rivalry on which the huge enterprises were based.

Starting just two years after the launching of *Sputnik 1*, repeated offers of co-operation were made by Soviet and American leaders. At the United Nations in October 1959, the Russians proposed an international conference of scientists on the exchange of experience in the study of outer space. The Americans were unenthusiastic at that stage, because the Russians, with their prevailing ascendancy in space, were expected to dominate such a conference. The conference finally took place nine years later, at Vienna in 1968, by which time the Americans had more to offer of value to developing countries.

From his 1961 inaugural address onwards, John Kennedy pressed the theme of exploring space together with the Russians. When John Glenn made the first successful American orbital manned flight in 1962, Nikita Khrushchev sent a telegram suggesting that the two countries should pool their efforts—'This would be very beneficial to the advance of science and would be acclaimed by all people who would like to see all scientific achievement benefit man.' Kennedy immediately replied, 'I am instructing the appropriate officers of this government to prepare new and concrete proposals.'

But the US National Aeronautics and Space Administration was unenthusiastic, and the eventual proposals were relatively minor. Eugene Skolnikoff (1967) described how, as a result of the coolness shown by NASA and the State Department in 1962, Kennedy's next effort was undertaken without consultation with those agencies. Shortly before his assassination, in an address to the United Nations, he said:

'Why, therefore, should man's first flight to the Moon be a matter of national competition? Why should the United States and the Soviet Union, in preparing for such expeditions, become involved in immense duplications of research, construction and expenditure? Surely we should explore whether the scientists and astronauts of our two countries—indeed of all the world—cannot work together in the conquest of space, sending some day in this decade to the Moon, not the representatives of a single nation, but the representatives of all humanity.'

The space adventure was not entirely a race between the USA and the USSR. Behind the headlines about spectacular 'firsts' a remarkable flowering of international co-operation involved many countries, primarily with the USA, in space research and applications. The USA provided launchers for the experiments of space research workers of many other nations, including several satellites, and room in its own satellites for their experiments on equal terms with American scientists.

Samples of lunar soil and rock that the *Apollo* astronauts brought back from the Moon were allocated, not only to Americans, but also to 27 foreign scientists for analysis. The development of transoceanic communications via satellite depended upon close collaboration between nations. The success of the US meteorological satellites was the trigger for the new global collaboration in all aspects of weather forecasting—the World Weather Watch.

The Soviet Union appeared less inclined to collaborate until de Gaulle's government came to an agreement for the launching of a French satellite by a Soviet rocket. On the other hand, Soviet research workers took a prominent part in the affairs of the international scientific Committee on Space Research (COSPAR) and its meetings. They reported and discussed their results; they disclosed their technology, however, only as it became obsolescent, and left foreign scientists worrying anxiously about whether or not they would conform with internationally agreed procedures for sterilizing spacecraft aimed at other planets.

As for collaboration with the USA, the Soviet mood in 1967, following Webb's appeal, was probably determined by the Vietnam War; at any rate, American astronauts were refused permission to attend Komarov's funeral. In the following year, the Russians released archive film of space launchings to the American NBC television company and, in an accompanying interview, Anatoly Blagonravov indicated an eagerness to collaborate with the Americans, if the political circumstances allowed. The UN Space Conference in Vienna in August 1968 offered an open forum and private rooms for further negotiations about collaboration.

An end to the space race would not mean a cessation of space activity; on the contrary, even if the budgets were cut sharply, research and useful technology in space might very well proceed faster, if some of the more spectacular flights for prestige purposes were scrapped. Even if there were no pressure to deploy in other ways some of the technological and financial resources committed to the space race, it would not be feasible for the Americans and Russians to continue extremely competitive activity in space indefinitely. In 1968, the American effort was visibly slackening, except in the remaining phases of the programme for sending men to the Moon.

The zone of space potentially accessible using current and foreseeable technologies extends much further—throughout the solar system and perhaps a little beyond. For the longer journeys, by unmanned probes, or for manned flights to the nearest planets (Venus and Mars) big advances in propulsion beyond the present chemical systems based on the German A4 rocket, would be called for. The cost and effort involved in a race to Mars, with target dates for manned flight there around 1985, would be immense. Without a deliberate decision unilaterally or by negotiation to end the race, it could continue—to a manned landing on Venus around 1990, then a manned landing on a satellite of Jupiter . . . and so on until an American or Russian (or Chinese) stepped on to the bitterly cold surface of the farthest planet, Pluto. The bored applause could then be drowned by the lamentations of an Earth stripped of its treasure for the least necessary journey imaginable.

Even if political barriers were to fall immediately, the development of a programme of collaborative or concerted operations in space would take many years to bear fruit. Both the Americans and Russians have committed large resources to the development of spacecraft and launchers, which blend with their own broad-based programmes of space exploration. Collaboration would therefore be concerned, in the first instance, more usefully with joint operations using the existing vehicles, or with avoiding unnecessary duplication.

The simplest form of collaboration in manned spaceflight would be in the tracking of one another's spacecraft. The Americans have a world-wide system of ground stations which the Russians lack. The Americans could enable the Russians to secure full benefit from their tracking network simply in exchange for advance information about impending flights, and the precise purpose and characteristic of flights in progress. Discussions along these lines could quickly lead to an exchange of programmes and timetables—and that could be the most important step in easing the race. The present asymmetry of openness and secrecy on two sides leaves the Americans having to assume that the Russians are, perhaps, smarter than they really are. Given an exchange of programmes, consideration of dovetailing of launchings and matching of programmes for the most productive scheme could follow automatically; when one side saw that the other would secure priority with a particular series of launchings for a particular end, it might scrap its corresponding programme. And then, most important of all, would come discussion about the overall timetable, and whether the two sides would agree to take their time and stop trying to better one another.

More positive possibilities begin with free exchange of experience of operating spacecraft, and the institution of a rescue programme whereby spacecraft of both sides would be kept in reserve as 'lifeboats' for astronauts in distress, and lead up to rendezvous in space in which the two sides, in orbit or at a lunar base, would bring equipment to be literally welded together to make an international observatory. Such a procedure bypasses habitual Soviet secrecy about ground facilities and launchers.

Given a trend of this kind, other nations will presumably be able to join in as they please. If the day ever comes for planning a manned flight to Mars with an international launcher and an international crew, non-competitive thinking may cast such doubt on the usefulness of so extravagant an expedition that it will not be undertaken in this century.

THE SPACE SALESMEN

The speed with which American engineers made 'pictures by satellite' a reality surprised the rest of the world, including the telecommunications and broadcasting authorities. But the communications satellites introduced into service in the 1960s relayed telephone conversations and television programmes simply from ground station to ground station—these stations being elaborate affairs costing several million dollars. The services were thus, at the start, firmly under the control of each nation's telecommunications authorities because conventional communications networks were needed to carry the signals to subscribers or audiences.

As the design qualities and transmitter power of the satellites increase, other possibilities arise. The size and complexity of the aerials needed to pick up the satellites can be reduced, first to something appropriate to a town or village and eventually to the point where each house or block can have its own 'ground station'. The direct-broadcasting satellite, analogous to a local television transmitter, will bathe a third of the inhabited world with its signals. Through such a system, vast exchanges of information will eventually occur including, for example, the distribution of newspapers by radio signal to printers in each home. For poor countries so far unequipped with networks of television transmitters the direct-broadcasting systems, or intermediate systems, could provide a short cut to national television services.

The belt, 22,000 miles above the Earth where the 'stationary' satellites hover, will enclose big issues of international politics, about the content of the transmissions, about the ownership of the satellites and the sponsorship of the programmes, and about the dominance by the technology, ideology and culture of the rich nations likely to occur through such channels.

Alva Myrdal (1967), Swedish minister of disarmament, warned: 'The technical possibilities to develop the telecommunications satellite to such a point that the most isolated village in the most distant continent can be reached by one and the same powerful transmitter, faces us with the choice of utilizing this advance either for establishing a system of equitable interdependence between nations and peoples, or for a system of as yet unimaginable cultural hegemony by the technologically, industrially and economically strongest nation.'

By no means far-fetched is the prospect that the Russians might put up a direct-broadcasting satellite over Singapore to beam propaganda at the Chinese; or that a commercial American satellite over the Maldive Islands would flood the Indian subcontinent with westerns and quiz shows. The 'pirate' broadcasting stations that sprang up on European coastlines in the 1960s would seem as nothing compared with a powerful broadcasting satellite. The diversity of mother tongues (fourteen recognized languages within India alone) and the cost of the receiving apparatus provide some substantial operating obstacles for space broadcasting pirates. Nevertheless, the threat of malicious or, more probably, unwise use of broadcasting satellites is such that some form of international convention is urgently needed to govern their uses. The issues are complicated because satellites cannot carry unlimited numbers of services without mutual interference: questions of allocation will arise.

UNESCO began examining, in the late 1960s, the practical possibilities of using satellites as an aid to education in the poor countries. A report by Wilbur Schramm (1968) looked soberly at the technological prospects, and did not see satellites as a simple solution to a complex problem—certainly not a universal solution. He wanted interested countries to begin thinking about how satellites might be matched to educational needs and teaching methods, and to stage some ground-based pilot experiments. Radio and television already help teaching in the poor countries. To make economic sense, educational satellite broadcasts would have to serve a very large area and carry material of common usefulness for a big region or group of countries. Thousands of field workers would be needed for liaison and feedback between the studio teacher and the classroom teacher. The effort might often be better devoted to conventional teaching. But the power and novelty of the broadcasting satellite should not be understated. Given satellites and non-dictatorial assistance in programme preparation from the rich countries, together with a rapid development of visual teaching methods making as little use of language as possible, they could provide a desperately needed means of overcoming the world's shortage of skilled teachers. It is a chance well worth taking, and even just to try it would help induce solutions to the broader international control of space communications and broadcasting.

Without such control national sovereignty will be directly

challenged by a proliferation of unilateral or, at best, multilateral activities dominating the channels and making their own rules. It is hard to see how international friction can be avoided when the mutually insulated cultures of East and West collide in orbit. Even if it is agreed that the aim is primarily educational, that only narrows the area of dispute: what is education in Kharkov seems like propaganda in Pittsburgh, and vice versa.

In the USA, the problem shades off into the politics of domestic broadcasting. As far as close co-operation with the Russians was concerned, that possibility waned when the private Communications Satellite Corporation (COMSAT) was launched on Wall Street, by Act of Congress, as the entrepreneur for international communications satellites, with half the shares allocated to the telecommunications industry. Kennedy had followed Eisenhower in deciding that international communications satellite operations should be developed by private enterprise; in a proviso that did more credit to his goodwill than to his practical consistency, he envisaged the UN playing a part in space communications. The agent for the USA in international dealings was to be a private corporation, and the policy was to be assertion of rights based on technical ascendancy. 'Who is there first has a priority, so to speak,' declared the president of COMSAT, Joseph Charyk (1968).

The picture is not entirely of that kind. The future shape of the International Space Communications Consortium (INTELSAT), using existing types of point-to-point satellites, had to be settled in 1969. Lyndon Johnson indicated that he would be willing to see COMSAT give up its majority holding in INTELSAT, which runs affairs for the Western world at present; he would also like Eastern Europe, including the Soviet Union, to take part in the future global satellite network.

In 1968, the Russians advertised their Intersputnik system as a rival to INTELSAT, offering 'one nation, one vote' in its management. But it was pie in the sky at that time, as the Russians had not even demonstrated communications satellites of the necessary type. At the INTELSAT meetings in Washington in 1969, civil servants from eighty countries had tough arguments about carving up the rights and profits in the new and potentially extremely profitable business. They sought to end the total American domination by transferring management of the system from COMSAT to a new international organization. The technological power of the USA was restrained by the customer power of its partners in INTELSAT: it takes two to communicate, and without agreement the satellites could be idle.

American television broadcasting is among the worst in the world, because of the commercial basis of operations and the intolerable interruptions of advertising. The Ford Foundation saw a great opportunity, in the communications satellite; in 1966 it proposed a scheme for quickly setting up a new domestic television network, to be financed from the proceeds of an associated communications system and therefore to be 'non-commercial'. But while some Americans wished to reform their home television services and leave out the advertising, others looked forward to the day when their world-wide salesmanship could be reinforced by world-wide television commercials. By the 1967 outer-space treaty, each state is generally responsible for the space activities of its nationals, but there is nothing in the treaty about the content of television broadcasts.

The first routine use of communications satellites was by the US Defense Department, employing a system quite distinct from that of the civilian operators to reach mobile ground stations in Vietnam and elsewhere. The first use of communications satellites for internal networking of television programmes began in the USSR. The first direct-broadcasting satellite has yet to be developed. There may still be time for political action to prevent a communications race which could turn outer space into a new Babel and the world into a cultural dustbin for the Lucy Show and socialist realism. Such defensive action could, if it were successful, convert one of the gravest threats to international goodwill into the most potent instrument yet conceived for the mutual enlightenment of the human species. Otherwise, we may all want anti-satellite missiles to switch the programmes off.

CHAPTER 11

Machines to Outwit Us

THE IDEA of the national computer grid passed out of the realm of science fiction and engineering speculation, and into open politics, in London in the spring of 1967. It might not have done so; it might have remained a purely administrative move. The British government was the first to make provision for a truly national, and nation-wide, computing service, with a constellation of large computers dotted about the country and available to all comers as a public utility. Logically, as the nation's biggest user of computers and also the monopoly owner of the network of telephone and telegraph lines, the Post Office was to provide the service—in the first place from spare capacity of a new massive order of big machines for its regional centres and for its Giro banking scheme.

The Post Office was to sell computer time, either under long-term contract or on an *ad hoc* basis; it was also required to help existing computer users and to provide channels of communication between computers. The intention was to develop a full national service as quickly as possible. Given the necessary will on the part of the Post Office, to provide a properly inter-connected network and to upgrade the machines to permit time-sharing from a large number of users with their own remote terminals, via the telephone lines, and given also the development of big storage banks for data, it was easy to envisage a quick advance towards a total information system. It would be a development at least as full of consequence for the nation as the building of the railways had been in the nineteenth century.

Yet, as I was astonished to discover at the time, the government expected that the necessary legislation to set up such a service would be non-controversial. Was not a system that would shape the character of Britain in the decades ahead worth a few arguments in the House of Commons? Accordingly, I wrote an article in the *New Statesman* (Calder, 1967c). It deliberately appealed to the party political instincts of the Conservative opposition, pointing out that only the innocent could suppose that we were not witnessing the nationalization of the most important service industry of the future. Although such a system seemed thoroughly desirable, provided the necessary safeguards were made, it was important to have it contested and for Parliament to warn the technical planners that it was watching them.

The article had some effect, and the opposition gave the government a much rougher passage for the Bill than it envisaged—sufficient at any rate for me to be rebuked by a minister for causing trouble. Crucial matters, at the end of the day, were covered only by declarations of honest intention by ministers, rather than by explicit legislation (the Act is a very simple one). Nevertheless, Parliament had not entirely failed to air the issues. As Jeremy Bray, parliamentary secretary to the Ministry of Technology, observed piously in the debate: 'I trust that the House will continue to exercise this vigilance not only in the discussion of the future stages of the Bill, but also in the surveillance of the system as it is discussed and developed technically.'

LIKE NOTHING SINCE WRITING

'Large-scale integration' was the watchword in the electronics industry in the late 1960s. The phrase meant, not a monopolistic tendency in the industry (though some saw that could be implied, too) but, technically, the manufacture of an extremely complicated circuit, too fine to be clear to the naked eye, on a single small slice of silicon. The thousands of microscopic electronic components could be duly tested and connected together in a working circuit, without the intervention of the girl with the soldering iron.

Miniaturization for its own sake was no longer the main aim. Outsiders might be impressed by the fact that the delicacy of some operations was equivalent to printing the complete works of Shakespeare on a postage stamp. But little circuits needed to save weight in missiles and spacecraft were already available. The transistor had made possible the construction of powerful compact computers which, using old-fashioned valves, would have required an opera house to contain them and would have been permanently under repair. The computer engineers wanted to go smaller still because, as they planned machines to work in billionths of a second rather than millionths, they had to allow for the fact that even an electric signal can travel only a foot in a billionth of a second. But the greatest attraction of large-scale integrated circuits was the promise of low cost and high reliability.

The exploitation of subtle electric, magnetic and optical properties of special materials, in the style of the new electronics, flowed from the big post-war effort in solid-state physics. It paid off in a range of novel or improved devices and techniques for manipulating signals of all kinds. Computers, the most important of the electronics industry's products, came into the manufacturing process itself, to design circuits—even other computers—and to supervise their manufacture, down to the introduction of the carefully controlled impurities on which many of these devices depend. The electronics engineer looked forward to being able to tell a computer roughly what was wanted and having the circuit designed and manufactured without further human intervention. Computers would thus, in a literal sense, breed new computers.

Future historians will probably be able to match the importance of the invention of the computer with nothing that preceded it, back to the invention of writing; even the invention of the steam engine and other heat engines will seem like a runner-up. To say that, just as the steam engine and its mechanical successors liberated men from muscular drudgery, so the computer is liberating them from mental drudgery, is not merely a cliché, but one that misses the point. The really significant thing about the steam engine was that it enabled men to accomplish physical feats, such as high speed of travel, of which they were previously incapable. The computer opens pathways to achievements in the realm of information and ideas previously denied to us. Its political and social implications are already staggering.

The steam engine was a feeble thing until James Watt introduced the separate condenser; similarly, calculating machines from the abacus onwards were unimpressive until they were provided with a store for numbers and instructions, as envisaged by Charles Babbage in the nineteenth century. He designed what we would now call a computer but he was unable to finish it with the mechanical devices available to him in Victorian times. Although the development of the computer has been intimately bound up with progress in electronics, that is a matter of technical opportunity. In principle, Babbage's mechanical computer could have worked and in practice fluid computers, which work by liquid or gas, are used today for some purposes. A computer is really just a machine that does simple arithmetic on numbers supplied to it, in accordance with a programme of instructions stored in advance in the machine. That simple description, though accurate, understates the enormous power of the computer and focuses too much attention on mere numeration; so many operations, logical systems and ideas can be converted into digital and arithmetical form that it is hard to see any inherent limit to its application.

A computer can run a business more efficiently than human beings could attempt unaided, using a working 'model' of the business in mathematical form and providing the chance to predict the results of alternative decisions before action is taken. In a factory, the computer wholly masters the arithmetic of stocks and orders and work loads, which has hitherto been largely a matter of guesswork. In most engineering design, including computer design as already mentioned, it liberates the engineer from detailed calculations and allows him to try out, imaginatively, many possible variations of the object or system under design. Much recent scientific research, including (to give one recent example) finding out the structure of the big key molecules of life, would have been impossible without the assistance of computers. And that is only a beginning. For the first twenty years, military uses, particularly in nuclear weapons research and in air-defence control systems, made the biggest demands on computing power. But now civilian tasks, especially weather forecasting, are beginning to make even greater demands.

A London catering company, Lyons, was the first to use computers in business and British industry has since pioneered new techniques and new uses for computers even though general application has been slow to follow. The operational research team of British Petroleum described the company's work in a series of 800 equations and so enabled the company to carry out production of oil from wells, transport, refinery operations and marketing, in such a way as to maximize profit in every step. BP's crude came from fifteen different countries with chemical characteristics peculiar to each field; it was transported in three hundred tankers, processed in forty refineries and marketed in seventy-five countries. So efficient was the computerized system in paring down the requirements for tanker bottoms that the company was particularly badly hit by the sudden closing of the Suez Canal in the Arab-Israeli war of 1967.

Most people will probably have their first experience of a computer by conversing with it by an electric typewriter. Among a series of important innovations in computers in the mid-1960s were the 'time-sharing' and the 'conversational' modes of operation which together allow anyone to work a computer without specialist intervention, and from his own office.

The first developments in time-sharing in computers were for special purposes, notably the American SAGE air defence system, in the 1950s. The idea was that, instead of confining itself to one task in one period, a computer should be able to jump about from one task to another in very quick succession, in response to human demands; thus several individuals could communicate independently with the machine, as and when they chose, and with a very fast computer the response would seem to be instantaneous. The first public proposal for time-sharing on a large general-purpose computer was made by a British expert, Christopher Strachey, at a UNESCO congress in 1959; in the USA John McCarthy had the same idea independently at the MIT Computation Center, at about the same time. By the mid-1960s, several laboratories had demonstrated that the necessary complex techniques could indeed be mastered, although there were some misgivings about the complications of handling very large numbers of 'subscribers' in a single system.

The basic idea of 'conversational mode' is that the computer asks the user what he wants to do and tells him what information it needs in order to proceed. The conversation is mostly in plain language, and so no special training is needed before you use the machine. When combined with time-sharing, this advance means that operation of a computer can be much more relaxed, casual and fitting in entirely with the user's own convenience. Men and machines can work much more effectively by such continual 'talk-back' than by the more conventional technique of shovelling programmes into machines like coal into a furnace.

A school of thought among a minority of computer makers has held that everyone should have his own small personal computer which, exploiting the extraordinary miniaturization now possible, he can carry in his pocket or keep in the drawer of his desk. Such devices will be made and will have limited local uses. But the way ahead almost certainly lies with large computer systems, linked to the users with the necessary communications channels. The natural, current development is the grid or network of computers, analogous to electrical supply systems. Huge 'data banks' for recording information of all kinds will be built into the network.

TOTAL INFORMATION

Computing power will come in much the same way that water, gas, electricity and telephones are provided today. The businessman will use the computer (which may be many miles away) as a scribbling pad and diary, as well as for much more elaborate operations for his company; the housewife will use it for doing her shopping, with goods displayed on a television screen; news, banking and a wide variety of other services affecting every walk of life will be conveyed on the same system. Lawyers will turn up precedents; doctors will record or analyze information about patients; children and their parents will play new games or plan a holiday, all on the same computing system.

Indeed, the only way fully to grasp the potentiality of combining computers with modern communications, and to judge what part they will play in daily life and in the work of the community, is to think of a system incorporating the computing, publishing, newspaper, broadcasting, library, telephone and postal services of the country, together with large slices of teaching, of government, of industrial and commercial operations, and of many professional activities. All these, each growing in its own right and subsumed in one system (or a small number of alternative systems) will together outstrip in magnitude and importance any industry or collective activity in which human beings have previously been engaged. If anyone asks where the capital will come from to create such a system, the answer is that it is money that the component undertakings would be spending anyway for their own computing and communications facilities, but now channelled into a common enterprise.

The purpose of any such computing and communications service will be to help organizations and individual human beings in their work, study and leisure. Everyone will have, in his office or in his living room, a console connecting him to the system. It is hard to forecast exactly how the various bits of hardware will be combined, and what programmes, material and storage retrieval facilities will be available through the system. But the console, or World Box, will probably combine telephone and teletype with a television pick-up and display—with these features greatly enhanced by the fact that they are feeding into, and fed from, a communal computing and storage system of great capacity.

It will be a total information service. The user will have immediate access to transitory information such as world news and stock market prices and so on, and also to recorded information from the world's books, periodicals and manuscripts in literal or facsimile form. He will presumably be charged according to the use he makes of the many services available, including small fees for the use of copyright books, films and so on. As banking will be handled in the same system, the user's account can be automatically debited and the publisher's, producer's or author's credited.

What will be the scope for individual choice within the system? Technical opportunity for great freedom of choice in the minute-by-minute operation will exist for the World Box users. Will this freedom be matched by scope for considerable individual participation and initiative in the provision of stock material and programmes for the system? Regardless of any labels of state or private enterprise, will service industries, lawyers, authors, entertainers, publishers, teachers and other professionals, be able to feed material in and advertise its existence to potential users? Will political propaganda be censored or rationed? Will the World Box be a vehicle for pornography, or will there be some latter-day Lord Chamberlain or Hays Office to sniff it out? In other words, who will control the system and what will the rules be? Decisions on these points will affect the character of the available information. Even if they are not government hacks, the men organizing the content on the system will be immensely influential.

This glimpse of a total information system capable of delivering millions of words and pictures into the living room perhaps makes that self-evident. The work of commercial and professional organizations will be transformed. There may then be no very clear distinction between authors, scholars, publishers, librarians, television producers or anyone else who can be called an information mediator—but it will be their task to save mankind from drowning in its own information, with its passages blocked by facts and opinions unlimited and available in any and every chosen form. Because the rate at which the human being can assimilate information via his senses and his brain is strictly limited, the result of an overload of information is confusion, distraction and eclipse of thought; alternatively information presented as painlessly and agreeably as possible could become a kind of drug. The system itself will have to provide selection, coherence, emphasis, shock, surprise, provocation, argument and aesthetic stimulus. The need for creative writing and editing will be greater than ever, even if quite often the form will be a programme or package rather than conventional prose.

Something equivalent to advertising may be essential, because even the most independently minded subscriber will be quite unable to discover what he wants in the electronic store without some guidance. At first sight it might appear that a great benefit of the system would be to break the domination, based on large financial investment, of big companies in the mass-communication field; in theory any individual could publish a book or article merely by typing it or having it automatically read into the system. But the trouble in drawing it to the attention of subscribers redresses the situation. Presumably the would-be advertiser must pay for transmission of information either into an index which subscribers would consult or to a stated list of subscribers (this might be the entire network) either during a period reserved for advertising or by some other means. The power and importance of advertisers, like that of editors, can only tend to increase—unless advertising is confined to information requested by the subscriber. It is a question of who is in charge of information about the information.

My anxiety is not that I may have been overstating the technical and organizational possibilities of the World Box but rather the reverse. It seems a foregone conclusion that such a total information system will emerge—not just because it is technically almost feasible today, but because people will genuinely want the extraordinary services it will provide. If we are not to be caught napping, we should look beyond these obvious possibilities to things more 'far-out'. I have not mentioned, for example, the extension of the programmed book idea to the reader-programmed books and so to the reader-programmed novel or dramatic plot—choose your own ending. Nor the likelihood that in the well-educated, leisured world of the future the system will have to accommodate vast and possibly desirable outpourings of amateur literature and works of art. To speculate that the World Box which links us to the total information system may eventually be something that we can carry in our pockets or on our watch strap—or in our hat, feeding directly into our brains and not via our senses—is probably no more wild than was the talk of space flight thirty years ago.

CENTRALIZATION OR DECENTRALIZATION

The World Box will alter everyone's lives in considerable detail, and certainly affect the style of daily work in all professions. For a start, the physical decentralization to the suburbs, made possible by the development of the railways and the automobiles, will be extended. Linked to his office and his clients via his console at home, there is no reason why a man could not be located anywhere and yet still be able to do his normal work. The development of such systems may have a big effect on the distribution of people across the face of the countryside, and help to restrain the growth of cities. A lot of extrapolations that assume a big growth in travel and transport of all kinds may be falsified. It should prove very convenient to carry on television conversations with one's friends and with the people with whom one does business, and to arrange video-conferences in which individuals in many parts of the country can be united face-to-face through their World Boxes.

Politically, the most fascinating aspect of the computerized information network is that it can give a powerful boost to centralization of authority—or to decentralization. Like a new road that can bring fresh life to a remote area or merely make it easier for the local population to leave, the information network is a two-way system. For example, it can either encourage offices to congregate in the metropolis, because they can keep in touch so easily with what is going on at the periphery; or it can encourage them to disperse because any place that can tap a comprehensive system is as good as another. That ambiguity shows us that geographical location will be an imperfect sign of the centralization of power. Rather, the question will be: Is there a powerful government group, perhaps diffused geographically, distinguished by its privileged access to information denied to 'peripheral' organizations, communities or individuals?

Information coming to the 'central' group from the 'periphery' can be used to enhance or frustrate the democratic process—either to serve as an indication of respected peripheral opinion, or to provide a handle for manipulating or censoring that opinion. Information flowing the other way can take the form either of orders or of comprehensive assistance for peripheral decision-making.

All forms of transport and communication are a potential aid to central government. But they can operate the other way, too, making it easier to organize a revolution. More prosaically, with the aid of the telephone, advertising and good transport, a small company miles from nowhere can market its products all over the country.

This issue of centralization versus decentralization of political and economic power is perennial and fundamental in politics: it is manifest in arguments about federal versus state powers, of central parliamentary authority versus local government; of big companies run from the financial centres versus the small company from far afield. It has strong technological connotations.

When horses and sailing ships were the best means of moving about and sending messages, government from the centre was not easy to maintain but, conversely, it was rather difficult for the outlying communities to have much influence on what went on in the capital. Commanders and governors had to be given local authority and trusted; if they chose to raise a revolt from their outpost, it was hard to check before it had assumed military significance. On the other hand, the remote governor could find that decisions were taken in the capital which affected himself or his territory, without any consultation. The shortage of ports and roads gave sensitive points for control, for levying customs dues, for regulating trade, safeguarding monopolies and arresting undesirables.

The impact of the steam locomotive was tremendous. It cut weeks of travel to days, and days to hours. The electric telegraph cut the days or hours to milliseconds for the purposes of sending messages, and made for greater efficiency and urgency in administration. But what were their political effects—centralizing or decentralizing?

Generally, the tendencies were on the side of the government. The most spectacular development was in the United States, where the huge railway networks united the country and turned the western frontier into a civilized part of the nation. The trans-continental railroads in particular, from 1869 onwards, played a special part in the growth of the United States. Thousands of miles of track were built with the assistance of land grants from the federal government, in return for which government traffic was carried at reduced rates for the greater part of the ensuing century. Moreover, the very importance of the railways in the national life made them a magnet for government control. The Interstate Commerce Act of 1887 brought the financing, fares and engineering standards of the railways under federal control; it was a big increment to the power of the federal government.

In other countries the central government took a more direct part in the creation and running of railways. Large parts of the Russian railway network were state-owned before the revolution of 1917, and even where the initial development of railways was by private enterprise, they have often been nationalized (Germany 1920, France 1938 and Britain 1948). Moreover, rapid and cheap distribution of goods by rail favoured the growth of large commercial organizations and monopolies. A small manufacturer somewhere far out along a railway line could use the network to establish his own business, but inevitably the tendency was towards the growth of large companies.

At first sight the automobile represented the very opposite of the railway. Its invention appeared to be—and in its early days perhaps was—a great victory for the individual. A man who could not set up his own railway might now buy a truck and carry his goods as he pleased. The individual could own his own car and travel 'freely', provided he could afford it; if not, he was better off with the centrally run railway. The automobile had a dispersing effect in the geographical sense. Just as the railways had encouraged the sprawl of suburbs along railway lines, so the automobile made it possible to build suburban houses in the intervening wedges where only roads would serve. There are areas, particularly in the United States, where it is impossible to live a normal life without two cars in the family. But we should not confuse the sprawl of cities and the growth of suburbs with decentralization. On the contrary, they represent growth around focal points.

In any case, economically and politically the decentralization was illusory. The roads had to be provided by public authority, and the building of fast highways became a preoccupation of governments. Elaborate law-making governed the use of the roads. The manufacture of automobiles became a major activity in most industrialized countries and almost an index of prosperity; great oil and rubber companies grew up to serve them. The emphasis on mass production of automobiles, from the time of Henry Ford, favoured the growth of very powerful industrial enterprises, which came to be identified with national well-being ('Buy, buy, buy—it's your patriotic duty'). The governments were very loath to interfere with their activities or to regulate or control the manufacture and use of automobiles. The community paid a very heavy price.

By the late 1960s, the free-ranging automobile looked obsolescent in advanced countries because of the congestion within cities and the prospect of new forms of surface transport operating at 300 miles an hour. Engineering solutions combining the convenience of the free-ranging automobile with the discipline of the railway seemed likely to supervene. Even Americans recognized that such systems would typically be owned by public authorities.

The importance of traditional estuary ports and inland communications centres was challenged during the industrial revolution by the growth of industrial centres in places determined by geology and in particular by the deposits of coal and iron. Locally, these centres drew former farm workers into their mines and factories; nationally it was decentralization, with economic power shifting out to the areas of chimneys and furnaces. Geology and technology also created the industrial proletariat, with workers crowded together in harsh cities and discovering a unity that peasants could never acquire. It was symbolic that Lenin went by railway to Petrograd, Russia's industrial capital as well as its political capital.

Today, more and more industry has shifted away from the transitional centres and derives its energy from electricity, and more and more people are finding their ways into offices rather than workshops. Paradoxically, power lines carrying electricity to anywhere, tend to encourage centralization because constraints of geology on energy supplies are negated and industries move to favoured centres of population. People and power thus shift back to the traditional or new centres of administration, communication and culture.

In the era of the computer network and the high-speed train, of the hot line, communications satellites and jet air travel, we can envisage systems in which instructions from the centre and opinion from the periphery have equal potency. In practice, control of the networks and transport systems will itself be the prime means by which general political control is exercised—as it was in the days of the horse and the sailing ship. The channels may become politically more significant than the senders or receivers of the messages, even as the significance of the message grows with the augmentation by the computer. The tendency depends on the system—both on the technology and on the organization. If a 'common carrier' can discriminate against particular users, or if the state or the wealthy advertisers control a broadcasting service, power accrues to the centre. The more expensive the system is to create—and computers and communications satellites are expensive items—the less avoidable is centralized ownership.

Technological development of the present general kind occurring in industry promotes centralization in another, indirect way. As traditional patterns of raw materials and products are abandoned, industries and companies become less secure and self-sufficient. They try to compensate by broadening their areas of activities, but they cannot help but rely on suppliers over whom they have no control; therefore, they have a stronger interest in seeing the business and the economy as a whole well planned at the centre.

Anne Carter (1966),an American social scientist, described the process as follows:

> 'The diversification of materials breaks down the primary identity of major industrial blocks. The increase in general inputs that render the same services and deliver such indistinguishable products as kilowatt-hours to all customers makes input structures more alike. But as a principal consequence of technological change the diverse major industries in the US economy tend to become interlocked in increasing interdependence. In the job market there is declining demand for people in the productive functions, as traditionally defined, and increasing demand for people who can contribute to the co-ordinative and integrative functions required by the larger and more complex system.'

Large size and centralization of power go together. Large technical or industrial projects of any kind, but especially where big injections of public funds for research and development are needed; large automated transfer production lines; large computers; large power units; all these are tending to create more centralized economies. The economics of industry encourages this tendency because in many processes, particularly those involving inherently elaborate plant, the advantages of scale are notorious. Conversely, those who favour decentralization or industrial competition or, in its traditional political sense, private enterprise, should favour small technical projects, small power units, small computers and versatile production methods.

The outstanding example of centralization within central government itself was the reorganization of the US Department of Defense under Robert McNamara in the 1960s. It rationalized and streamlined the operations of the American defence machine, and suppressed inter-service rivalry. It was motivated by two great fears of the late 1950s: Soviet advances in rocketry (the alleged 'missile gap') and the apparent requirement for mobile conventional forces (as in the Lebanon). But as the years passed it became plain that the centralization oriented all the military programmes to one man's view of the future, and there was anxiety about whether McNamara could be successfully replaced in the exacting role in which he had cast himself (Encke, 1967).

A further instance, showing effects on provincial government: by tradition Sweden has powerful regional and local administrations, but advances in technology and associated problems are forcing a change. For example, a multitude of governing bodies for education and health services is not compatible with the wish to develop, nationally, advanced technologies in education and medicine. Again, problems of water pollution and air pollution are difficult to cope with while each district has jurisdiction over the location of industry, and may accept noxious plant against the wishes of the conservationists and the national government. Slowly Sweden is readjusting its political balance towards the centre.

When the world is encased in an electronic cocoon of interconnected computer networks, and more and more administration and business is international in scope, the issue of centralization versus decentralization will be a global one. The wish for universal law is strong, especially now that nations can annihilate one another; it does not follow that world government should be a centralized administrative machine. I foresee a time when the same people who now press for increasing world government will be defending the sovereignty and cultures of nations as a mode of decentralization in a global system.

TOOLS FOR TYRANTS

Even apart from questions of administration and ownership of a computer information system, the use that governments and other big organizations make of it will be a great source of central power. The national government can employ the system to gather information about individual citizens; it can, in principle at any rate, pool information collected for a variety of purposes—income tax returns, social security information, police records, medical data and so on, about an individual, gathered together to create a veritable dossier containing more personal information than it may be good for the state to have. It is from here that fears that the computer system may turn into Big Brother principally emanate, in the vision of huge government data banks in which the traditional requirements of privacy and secrecy concerning personal information in government hands are eroded. 'Whatever privacy remains to American citizens,' commented one senator, Edward Long (1968), 'it remains because the federal government is presently too inefficient to pull all its personal information files together.'

A US government plan, first mooted in 1965 by a committee under Carl Kaysen, would set up a central data bank (National Data Center) to pool information from a score of agencies. It provoked anxious discussion, and congressional hearings by Cornelius Gallagher's subcommittee in the House and Edward Long's in the Senate. Isaac Auerbach (1967) at the International Center for the Communication Arts and Sciences, New York, pointed out that the Fourth Amendment to the US Constitution protected the right of the people 'to be secure in their persons, houses, papers and effects, against unreasonable searches and seizures'; it had been interpreted in law against wiretapping and electronic eavesdropping. But Auerbach wanted something stronger. He proposed a Constitutional amendment providing 'that no person may be deprived of his privacy without due process of law', and giving people the right to know what information about themselves is in the files.

The argument in favour of much more elaborate, publicly controlled records about individuals is not a weak one. Greater and more detailed information about individuals can make for more efficient government. It may also, if properly handled, lead to more humane government and greater attention to the individual's wishes and needs. Social measures perforce treat individuals as undifferentiated units within a communal mass, or assign them to very broadly defined groups which take little account of idiosyncrasies and private circumstances.

Sociologists study individuals' behaviour, circumstances and opinions in great detail—and with great sympathy. But they are out to produce statistical tables from samples of respectable size; better still, to make generalizations about social problems or policies. To the extent that they help administrators to take account of broad patterns of human variability, they may improve life for neglected groups. There is no concealing most sociologists' implicit dedication to the idea of harmonious community. They, and their scientific and commercial associates, are themselves a threat to privacy. Taking opinion polls, visiting homes with questionnaires, making psychological investigations for any purpose other than the benefit of the subject himself—all these are intrusions of varying degrees of impudence.

Yet social research, census-taking, tax collection, public health inspections, criminal investigation and many other incursions into privacy are accepted as not merely sensible but desirable. The present issues evoked by data stores are neither wholly novel nor black-and-white. The result of agitation may, in the best outcome, not be a prohibition of all new developments in social administration and social sciences which could benefit from the extraction and computer storage of information from individuals. Rather there may be clear, legal definitions of the legitimate and illegitimate uses of that information, and limits beyond which probing is regarded as a nuisance or a danger.

Privacy is the individual's chief defence against regimentation or pressures to conformity. Walter Bagehot (1872) wrote: 'I remember at the Census of 1851 hearing a very sensible old lady say that the liberties of England were at an end; if Government might be thus inquisitorial, if they might ask who slept in your house, or what your age was, what, she argued, might they not ask and what might they not do?'

And in an article 'Don't Tell It to the Computer', the writer

Vance Packard (1967) commented sharply on the proposal for a National Data Center. He lodged objections on the basis of four major dangers, which can be paraphrased as follows:

People don't like being mere numbers in a machine.
People will distrust the government and may falsify information.
Computers don't forget, and a man may be a 'lawbreaker' for life as a result of a trivial childhood incident.
Information about an individual can be used by officials to intimidate him.

Such fears are well-founded and need to be expressed forcibly. It is no easy matter, technically, even given the will to make the necessary legislation and practical arrangements, to ensure that only those people authorized to have a particular bit of the information will be able to reach it. In principle, access would be by a password; but *someone* has to have access to the list of passwords. At the Massachusetts Institute of Technology, students once 'busted' the privacy arrangements by persuading the machine to print out all the details!

The technology of 'bugging' is another clear threat to privacy. It includes such facilities as conventional cameras, tape recorders, phone-tapping equipment, electronic stethoscopes, acoustic 'telescopes', X-rays for reading mail, and invisible but fluorescent markers for trailing individuals. Particularly ingenious is the 'harmonica bug' which turns an ordinary telephone into an unsuspected pick-up device. The incentives for the development of these technologies have lain in espionage, national and commercial, and in crime detection. Just as more efficient administration and social science in government provide the reason for the data bank, so the internal war on crime, and the external intelligence war, have given a kind of justification to bugging. Indeed, a conundrum for privacy-defenders is the following: what bugging devices would you rule out in a police effort to stop criminals gaining illegal access to personal files in an electronic store, with a view to blackmail? In 1967, however, the US attorney general Ramsey Clark banned wiretapping and eavesdropping by federal agents, except against foreign spies.

Police forces in various countries are turning increasingly to new technology. Helicopters and speed traps were early innovations. Nowadays, computers assist in traffic control, identification of stolen cars and turning up information on wanted suspects. There are hopes that a new laser photographic technique—holography—will assist in the very rapid identification of fingerprints. The possibilities go on and on. On unclear scientific evidence, 'voiceprints' have been much publicized as a means of identifying people by the characteristic frequency distribution in each voice.

'People have a right to do anything that's not forbidden by law, and there's no law against lying to you.' Thus General Peckem to Colonel Scheisskopf in Joseph Heller's *Catch-22*, in a fair summary of contemporary attitudes. But modern methods of communications, like radio and television, can carry the lie to a huge audience; modern methods of recording information, on microfilm or magnetic tape for example, can preserve the lie for posterity. To the saying 'liars need good memories', we can now add: 'and their victims have good storage'.

The most direct intervention of technology into the realm of human truth and falsehood is the so-called 'lie detector' or polygraph. It is simply a device for recording changes in a man's breathing, blood pressure and the electrical resistance of the skin of his hand, while he answers questions from an investigator.

The supposition is that the guilt or anxiety of a lie will reveal itself in involuntary departures from normal in one or more of these patterns. But in practice there is huge scope for misreading a polygraph. For example, if you give a truthful answer about something that worries you for other reasons it can register as a lie; or if you want to confuse the readings you can simulate a lie simply by tensing your toes or thinking of naked women. An American psychophysiologist, Burke Smith (1967), commented: '. . . to say or imply that the machine is infallible is to use a lie to detect a lie. To elicit admissions through fear of the machine or misrepresentation of its record is to force a confession.'

Polygraph operators argued that the crusade against them by psychologists relied largely on experiments with college students in fictitious crime situations. They claimed that the number of people cleared of suspicion and unjust accusation with the lie detector was much greater than the number found to be lying. The American Polygraph Association laid down standards of professional conduct. Nevertheless, by 1967, ten states of the USA had outlawed polygraphs for screening employees. Anti-polygraph laws in two other states (New York and Wisconsin) were vetoed by the governors.

Even if modern techniques used by the police are intrinsically sound, and powerful against criminals, the public will have misgivings about them. They will do so especially where relations between police and public are poor, or where the police have political functions. For the citizens in any existing 'police state' the prospect is extremely grim, in view of what electronic and other surveillance systems can do. In countries where politics is primarily a matter of detecting and imprisoning one's opponents, the potential impact of current technology on politics could not be greater.

Free countries cannot guarantee that they will always be free, and the existence of such potent public technologies as bugging and the electronic file may actually reduce their chances of so remaining. The thought of a Joseph McCarthy having access to a data bank is alarming. When comprehensive computer networks exist, reaching into every office and home, and people become dependent on them, it might be simply too difficult to give up using them if a dictator came to office; and, of course, spurning them might itself be regarded as a suspicious act. Given a fully operational network the dictator would have a very easy life. Ritchie Calder (1968) proposed, not frivolously, that the data banks should be programmed to destroy themselves if a Hitler came to power.

Existing notorious techniques of 'brain-washing' may be less effective than fiction writers and the public believe. But the development of new drugs affecting mental performance or mood will almost certainly provide a chemistry of tyranny. Tear gas and the potential use of drugs similar to LSD for riot control, will look crude compared with possible additions to drinking water that would keep people docile and cheerful. Witnesses of a scene of political terror might have the memory of it erased from their minds by the release of a suitable gas, before permanent memory has been formed in the brain (antibiotics have been found to erase short-term memory in goldfish). Again, it would probably be not at all difficult to make the population as a whole addicted, unknowingly, to a particular drug added to the food supply, so that they would eventually behave in almost any prescribed way to maintain their supply of the drug. Much more dramatic and direct control of mental behaviour is possible with electrical stimulation of selected parts of the brain, using small wires implanted through the skull. Jose Delgado, a Spanish-born physiologist of Yale, has achieved remarkable control over the behaviour of animals in this way. The logistics, for a dictator, of implanting electrodes in every head are plainly extravagant, but the job might be done for him, in advance, by individuals in pursuit of the pleasurable sensations that implanted electrodes can give.

Possibilities for mind control are thus apparent at the outset of the present intensive phase of research aimed at discovering how the brain works. With greater knowledge, yet more possibilities may suggest themselves. Other implications of brain research are discussed in the next chapter.

In a quite different direction—overhead, in fact—is the threat to privacy out of doors that comes from the development of photographic survey methods using aircraft and satellites. These techniques, devised in the first instance for military reconnaissance, are being refined continuously and applied to more and more civilian purposes, ranging from map making to detection of forest fires. Their value is greatly increased by the use of multiple cameras that take simultaneous pictures of visible and invisible wavelengths and allow, for example, species of crops to be identified. From the altitude of a satellite, it is not possible to observe anything as small as a human being, but we must expect increasing overflights from aircraft for all kinds of useful photographic surveys, at much lower altitudes. The police will want to use them for traffic purposes and possibly for tracking criminals or stolen cars—also for keeping a brotherly eye on us all?

The reason for rehearsing such possibilities, for computer surveillance, bugging, mind control and aerial observation, is not to cause alarm or gloom about the future but merely to ensure that people are on their guard. A bare quarter of a century since Hitler died, present police states seem less technologically minded and less single-minded than his, but the risk of tyranny by modern means is ever present. Moreover it does no harm if fear of outright tyranny makes people attend to more insidious threats to privacy already existing, and pending, in freer societies.

Lest this catalogue should imply that technology is inherently anti-libertarian, it is as well to recall the events in Czechoslovakia immediately after the invasion by the Soviet Union and other countries, in August 1968. The free radio, kept going by quick footwork and improvisation, was the most important single factor in sustaining the morale and passive resistance of the Czechs and Slovaks. It relied on the technologies of tape recording, microwave links and the transistor radio.

INTELLIGENT MACHINES

'I know few significant questions of public policy which can safely be confided to computers,' John Kennedy (1963) remarked shortly before his death. Thereafter, notions of what computers will be capable of doing have grown less restrictive. A highly developed computing system or artificial intelligence will be very clever, surpassing human performance in respects other than mere computation, as judged by stringent tests. Indeed, no one will be able to grasp what is going on inside the machine, in order to monitor and criticize the results.

It is not yet worth losing sleep over the science-fiction vision of machines revolting against men. It may be some decades before men are able, let alone willing, to omit the 'Off' switch from the machines, or give them direct access to energy and raw materials. But it may not be legitimate much longer to regard the computer as a high-speed moron that can only do what it is told to do.

Research in 'artificial intelligence' aims at the development of special machines, or programmes for general-purpose machines, that will act as if they are intelligent—that is to say, will respond to test questions and problems in ways that, if humans so responded, we should judge them intelligent. Important faculties that have to be cultivated include the ability to learn by experience, to make rough guesses, and to select a sensible approach to a problem encountered for the first time.

There are good, practical reasons for such developments. Persuading computers to recognize letters and numerals has proved to be exceptionally tricky, as those funny numerals on cheques bear witness; similarly machine translation of texts has also failed to develop satisfactorily despite many years of work, because the machine cannot match human ways of thinking. And although mechanization of indexing and cataloguing in libraries is very effective at the routine level, the task of selecting sources in ways that librarians find easy tends to defeat the machine.

For these and other reasons, men are trying to build or programme cleverer machines; at first to be as bright as we are, then brighter. Not the most likely route, but a last resort perhaps, will come when men understand how the brain works. They will almost certainly be able then to make an electronic version on similar principles, with the same number of components, but working faster. Given success in this direction of intelligent—and hence 'creative'—computers, we must expect a super-intelligent machine, transcending our own intelligence, and able to digest mountains of information. Shall we bow before it? Symptoms of computer worship are already visible, though the machines are, so far, not intelligent at all. The crucial issue in political electronics around the end of the century will be simply this: shall we rate the conclusions of a very intelligent computer more highly than the opinions of mere men?

Just as nuclear energy forces us to clarify our ideas about international relations for ever, because the knowledge that thermonuclear bombs can be made cannot be erased, so does the computer oblige us to adjust our attitudes to the existence of machines that may become smarter than we are. It does nothing less than oblige us to reconsider the role of 'rationality' in human affairs, as compared with more 'human'—in a sense, more animal—qualities.

If men do not wish to leave decisions concerning human needs to the computer, they must set in competition with electronics their human blend of reason, emotion and irrationality—accepting the interactions of the human nervous system with the blood-stream and endocrine glands, with other individuals, and with the natural and electronic furniture. Only on such a basis can they draw lines to mark off those crucial activities over which their non-electronic brains must rule.

CHAPTER 12

Biological Conundrums

IN PRE-ARRANGING the sex of offspring, the first experiment that left no doubts about its possibility was done with rabbits in 1967–8, by two physiologists at the University of Cambridge. They thus achieved something that had teased biologists for many years. Previous efforts at sex control depended chiefly on the fact that male mammals, including rabbits and men, produce two kinds of spermatozoa, 'male' and 'female'. Surely, the biologists had thought, there must be a physical or chemical difference that could separate the 'male' and 'female' sperms—by sedimentation, or an electric field, or in some other way. They had no luck. Robert Edwards and Richard Gardner at Cambridge concentrated instead on the mother and on the possibility that very young male and female embryos might be distinguished before they had attached themselves to the mother by implanting themselves in the wall of the uterus.

A few days after fertilization, and before implantation, the cells of the female rabbit embryo could be unequivocally identified under the microscope by the presence of a characteristic feature called sex chromatin. To sex the rabbit embryos, Edwards and Gardner had first to remove each from its mother and then extract by 'microsurgery' a few hundred cells from the embryo, for testing, before replacing the embryo in the mother—or another doe rabbit. After this rather drastic treatment, many of the foetuses failed to develop but, in the first series, eighteen did. They all proved to be of the sex predicted from the cell test.

The technique, or something similar, would before long be extended to other mammals, including human beings. Some less drastic way of distinguishing the sex of the young embryo removed from the mother was sought. The possibility of discarding embryos of an unwanted sex existed. The animal breeder might benefit enormously from the ability to arrange that dairy cows, for example, produced only female calves. There was also a medical reason for hoping to apply the technique to human beings: some serious inherited disorders, such as haemophilia, occur mainly in boys, so that a mother who suspected she was a 'carrier' of a bad gene might wish to make sure that all her children were girls. Beyond that lay the much-discussed possibility of even genetically healthy people coming to pre-arrange the sex of their children, for sentimental reasons, for social reasons, perhaps even for political reasons. The supposed preference for boys seemed to some to promise a ready check to the growth of the population.

Before Edwards' and Gardner's experiments, it was already feasible to discover the sex of an older foetus, before birth, by cell-sampling, and to abort it if it was 'wrong', just as one could abort the foetus if its cells revealed some gross genetic error, such as mongolism. But in view of many countries' prevailing moral and legal sanctions against abortion, destroying the unwanted embryo before it linked itself to the mother seemed less murderous to some experts. It was supposed, in any case, that intra-uterine contraceptive devices (the Loop) probably worked by preventing the fertilized egg from attaching itself to the uterine wall.

Opportunities are arising for intervening in the processes of human life from conception to death, which provoke complex questions of public policy. If anyone has invented a convenient term for this unconventional medicine I have not encountered it: 'eugenics', 'foetology' and 'euphenics' cover big areas between them, yet exclude others, while 'biotechnology' has altogether too broad a meaning. I shall call it 'athanatics', from the Greek word *athanatos*, for immortality; it carries the twin connotations of god-like power for manipulation of people, and the brands of possible immortality intimated in some of the manipulations.

In 1968, the National Research Council in Washington set up a committee of social and behavioural scientists, biologists, physicians and lawyers for a three-year investigation of biology and social policy. The possibilities can perhaps best be summed up by saying that Aldous Huxley's forecasts in *Brave New World* (1932) of babies bred in bottles and of surrogates for everything, understate them. They will also shake almost every simple idea by which men and women have always lived. The 'facts of life' are no longer invariant.

As a pre-athanatical man, I have a father and a mother to whom I owe my life. I 'take after' them and their ancestors; in modern terms I inherit an assortment of genes from both parents. For nine months I grew from a blob of jelly to a recognizable human being inside my mother's body before the cataclysm of birth. Then I was cared for in the family group during the perilous years of childhood. Later the natural passions took hold of me and, with one woman, I produced children of my own. Society is organized on the assumption that that is how I shall act. In my daily life, my moods, my pleasures, my anxieties, are generated according to events and my state of health, and are only loosely under my control. I expect to die but until then I expect to remain consciously the same person. In life after death I do not believe, but I admit to fancying a diffuse immortality for my genes, through a long line of natural descendants.

Such are the entirely commonplace circumstances of my life, and of billions of other human beings, past and present. Yet it may be technically possible for my children to break the pattern in every particular and lead utterly different lives. Conception of new human beings—my legal grandchildren, for the sake of argument—will be separable from sexual activity, or birth, or both, in a variety of ways.

Already, of course, birth control techniques can prevent conception and hormone treatments encourage it, as a result of normal sexual behaviour. Already, too, artificial insemination of women, using either the husband's sperms or a donor's, is used to end sterility in some marriages. Some men make their living as sperm donors, though why anyone prepared to live that way should be regarded as a suitable father is hard to see. Critics are concerned about the risks of accidental incest and inbreeding following large-scale fatherhood by anonymous donors.

Some biologists, notably Hermann Muller and Julian Huxley, have suggested an extension of the donor technique into normal marriage so that a couple may decide that the wife should conceive by some admirable man, perhaps long since dead, whose sperms have been preserved by freezing in egg yolk and glycerine. If that proposal seems unfair on the husband, try the next one: that the woman, too, can relinquish her hereditary rights in the matter and have planted in her uterus the egg of another woman, pre-fertilized by the sperm of another man, and be given an accompanying prospectus describing the virtues of the true parents.

You may think yourself so admirable that, so far from abdicating to an admired donor, you would like to produce offspring exactly like yourself, with no confusion by throwbacks to your ancestors or mixing with your wife's genes. It should be possible artificially to prepare an 'egg' containing your body's genes, which will then grow, in a woman or elsewhere, into a new human being—reproduced by much the same principle as a gardener propagates plants from cuttings.

Why should a woman carry the child, if it is someone else's or even if it is her own? Zoologists in Oxford implanted a fertilized mouse egg in the scrotum of a male rat (wrong species, wrong sex, wrong sort of place) where it grew as a normal mouse embryo. The trick was not very difficult. A fertilized egg produces a remarkable material called trophoblast, whereby it 'takes root' in the uterus and creates conditions for the embryo to grow. Trophoblast invades the tissue of the uterus until it reaches a blood supply; but if the egg is transplanted it can do the same thing in any part of the body, or in the husband's body, or in another animal. If neither parent wants to bother to carry the child, why not put it in a cow for nine months? Naïvely I assumed that some maternal instinct in women would always lead them to resist such an 'unnatural' procedure, until I heard some English girl students say they fancied the idea.

Other possibilities, such as conceiving and raising human

embryos in bottles on a production line, may be technically a little more awkward. But whether the fertile human egg is in a woman, a cow or a jar, before birth it will be susceptible to tests—like those for sex and genetic abnormalities, already mentioned—and to manipulation. Even after conception, there is still scope for modification—perhaps by injection of virus-like vehicles of genetic material, DNA, that will 'infect' the embryo with additional genes. Drugs, hormones and nutrients may encourage not only healthy but 'super healthy' development of particular organs or limbs, in attempts to make brainier or more athletic children.

THE EUGENISTS

These possibilities take us directly into the complex issues of human 'improvement', or eugenics, by athanatical intervention in human mating or conception. It is useful to distinguish different modes of eugenics to which moral and political responses may vary greatly.

Negative eugenics is concerned with safeguarding the existing human genes against mutation and with preventing propagation of obviously bad genes. Being chemical compounds, genes are subject to change as a result of physical or chemical violation, making the equivalent of a misprint in a line of type. Such mutation can harm the original owner of the gene by causing cancer or, if the genetic misprint is transmitted, a child may suffer congenital defect of a fatal or non-fatal kind.

The natural causes of genetic mutation are probably legion. They include the cosmic radiation which bombards us from outer space and the natural radioactivity of the Earth's surface. Even the random movement of molecules due to the warmth of the body is sufficient to cause an occasional mutation. There may be mutagenic agents in the food that we eat and in the air that we breathe. Additional mutations are caused by man-made agents, including the radioactive fall-out from nuclear weapons tests, and X-rays and potent chemicals used indiscriminately. A first objective in negative eugenics is that of keeping to a minimum any exposure to agents likely to cause genetic abnormalities. There is a close parallel in prevention of other congenital diseases, such as those caused by illness or drugs in the mother during pregnancy.

Men liable to suffer higher radiation doses as a result of their professional work—radiologists, nuclear scientists and astronauts, for example—may in future bank frozen sperms in case of accident, so that their wives could subsequently have children without special genetic risk. Indeed, Bentley Glass has suggested that, because mutations accumulate in the germ cells with advancing age, all males at a certain age may put their sperms in a bank to preserve their genetic integrity. At present, however, the storage procedures themselves seem to give rise to mutations, so the eugenists are not yet winning technically on this point.

Bad genes already exist in the population. By counselling, legislation or therapy, negative eugenics can seek to contain or prevent their ill effects. Even in this apparently straightforward human exercise, caution and humility are needed. Although most genetic disorders seem entirely undesirable, there are exceptions. Sickle-cell anaemia is a fatal hereditary condition occurring in children both of whose parents carry a bad gene, with a 'misprint' that alters just one amino acid among the hundreds present in the haemoglobin molecule of blood. The gene is very common in some parts of the tropics. The reason why it persists, despite its obvious ill-effects in the unlucky offspring, is that an individual having the sickle cell gene from only one parent has almost complete immunity to malaria.

As knowledge of human genetics grows, it becomes possible to discover bad genes in healthy carriers, and to give advice or take precautions to safeguard the children of weak individuals. Glass (1966) forecast that, by the 1980s, 'genetic clinics for testing prospective brides and bridegrooms will be a regular feature of every first-rate hospital or health department, since by then the number of such detectable defects may reach a hundred or more'.

In many cases, there will be a fifty per cent chance of a child inheriting a given defect. Is the couple, aware that one of them carries a particularly serious defect, going to give up having their own children (unless by egg implantation or donor insemination) because the risk of a defective child seems too great? Or will they take a chance, perhaps aborting the embryo if tests show it to be unhealthy? Should the state allow people to reproduce who are known to carry dangerously defective genes—for example, individuals saved by therapy from the worst effects of their mutations—or should they be sterilized? Everyone carries some bad genes, so shall we make ourselves extinct rather than risk imperfect children?

Nature is a good 'negative eugenist'—foetuses with serious mutations tend to be aborted or still-born. Happily genetic material has some capacity for correcting defects that may be produced—although it cannot always be done. This comparatively recent discovery was made in bacteria. The mechanisms are not well understood, nor is it known what kinds of defects can be repaired. Nevertheless, it may prove possible to use the knowledge of the DNA repair mechanisms to improve, for example, the treatment of cancer, by reducing the cancer cells' ability to repair themselves after treatment with radiation; conversely, the repair of hereditary defects may be encourageable.

Positive eugenics is concerned less with prevention of bad genes than with encouragement of good gene combinations, by analogy with improvement of farm livestock by selective breeding. Artificial insemination or egg implantation may be socially more acceptable than, say, the direct mating of young Nazis with favoured German women during the Hitler regime. The arguments for positive eugenic programmes are not to be shrugged off lightly. Hermann Muller favoured parental selection, but he dissociated himself from the racists and from the old-style eugenists who thought that the upper classes should have the reproductive advantage. He proposed a modest beginning. Where women had to resort to artificial insemination from a donor in any case, and perhaps also for *some* children in a normal family, pioneers might be willing to choose sperms from fathers of 'outstanding gifts of mind, merits of disposition and character, or physical fitness'. Muller believed that these pioneers would produce such excellent children that other couples would follow.

Muller also argued against the existence of a true instinct of 'genetic proprietorship' on the part of the father. It was necessarily absent in peoples whose lack of knowledge prevented them connecting copulation and conception. Nevertheless, those who do know may not take kindly to a step-parent status, nor may the law look kindly on the practitioners or products of remote mating.

Propagandists for eugenics used to lean on apparently incontestable evidence that the larger the family of children, the less intelligent those children were. The idea grew up that people of low intelligence were reproducing themselves more effectively than the more intelligent, so that the average level of intelligence in the population must inexorably be diminishing. Sheldon and Elisabeth Reed (1965) of the University of Minnesota demonstrated that this was a fallacy. Comparisons from one generation to the next had been made on the basis of parents on the one hand and children on the other; people of the older generation who did not have children were excluded from the comparison. Yet, the Reeds showed that the unmarried people of the parents' generation included most of the people with very low IQ. People who are severely retarded mentally just tend not to get married. It must be admitted, however, that new procedures for minimizing retardation in someone with bad genes may encourage the reproduction of those genes.

Why bother to 'improve' human beings? Normal sexual reproduction is the means of repeatedly shuffling and permuting genes in continuous experimentation with new individuals, each one unique. Given such a range, given the latent powers never fully exploited, and given men for any human purpose—whether it be to teach theoretical physics, to fly a spacecraft or to create works of art—why single out and favour individuals with particular genes and combinations? Or will it be thought 'unfair' to let individuals be born dull and ugly, when we have the means to make them bright and beautiful?

If the merits of natural human diversity are deployed as an argument against the 'positive eugenists', the 'inventive eugenists' can turn it in their favour. They wish to use knowledge of the genetic material and its code to create novel genes, or improbable combinations of genes. Efforts in inventive eugenics ('genetic surgery' or 'genetic engineering') will be on a much more artificial level than, say, insemination by a donor. The biggest practical objection to attempting any such thing in the foreseeable future is that there are certain to be hideous errors, manifest in human beings deformed in body or mind.

Inventive genetics is not practicable with human beings at present, although it is already being practised in a crude manner with plants, by subjecting their seeds to penetrating radiation in order to induce mutations. Among many defective progeny an occasional variety with particularly desirable characteristics appears. Other, more subtle, possibilities are emerging, particularly in genetically dependent hybridization of plant species which normally do not mix.

'Geneticists are not ready to conquer the earth, either for good or evil.' So said Salvador Luria (1965) of MIT, in seeking to start an earnest debate on the genetic direction of human heredity. Despite his disclaimer, Luria, a pioneer student of the genetics of viruses, could envisage technical methods for altering genes, based on experience with bacteria.

One is by the introduction of specially prepared viruses to live and reproduce within the body, and find their way into the germ cells, which they will alter without damaging them. In some such cases, the genetic material of the viruses might actually become incorporated in the genetic material of the cells; alternatively, human genes from one individual might be tacked on to viruses, which might then provide a vehicle for transferring genes to another individual. Again, it may be possible to extract certain genetic material from cells and transfer it to recipient cells. Going further still, men may actually synthesize artificial genetic material for introduction into the recipient cells.

Another approach is to learn to 'read', and then to 're-write' or suppress, pre-existing genes. Antibodies—the materials produced to attack foreign material as part of the body's defences—might be tailor-made to attack particular genes. Other tricks may be played to exploit the fact that genetic material is copied sequentially, letter by letter, sentence by sentence. Interference at particular times or particular locations along the train of genetic material, could pinpoint the genes to be modified or deleted. Yet more possibilities may arise with components of the human hereditary system dissociated from the chromosomes—components called 'accessory genetic elements'. These may be added or deleted more readily than the genes on the chromosomes.

Luria commented about the possibilities of developing such genetic control: 'The required advances in technology may not be conceptually greater than those that were needed to convert nuclear fission from a laboratory experiment to a self-sustained chain reaction.'

What use would human beings want to make of powers of genetic manipulation? Certainly, one must expect ideas of supermen, or genial moronic slaves, to be fancied by some athanatists. American engineers have proposed the development of little men, to relieve problems of crowding! Another mild but insidious application, with wide appeal, might be cosmetic—with a view to producing some desired colour of eye or style of hair or, more radically, to alter hereditary skin colour towards some ideal, presumably white. (The Black Moslem eschatology holds, conversely, that white men were created by a wicked black geneticist six thousand years ago!)

MODES OF IMMORTALITY

Like Laurence Sterne getting Tristram Shandy's mother to parturition, I have devoted a lot of space to having the athanatical baby conceived and born. That is only the start of a long chain of medical and biological possibilities, continuing through life.

Beyond present efforts to replace organs by transplants or implants, the clever procedure in future may be to restore or reinforce organs, or develop novel ones, by adapting the natural processes of regeneration. Just as a lizard can grow a new tail, or a snail a new head, to replace one mislaid (or cut off by a zoologist) so a man may be able to grow a new heart. With suitable opening up of the skull to prevent overcrowding, he may be able to grow extra brain capacity, like a tumour on the top of his head. Or, if he wants to live an aquatic life, it may not be impossible to release the evolutionary memory in his genes, of how to grow a working pair of gills in his neck.

If the individual's consciousness and identity are chiefly represented by his brain, one mode of immortality is not hard to identify. The culmination of a policy of replacing defective organs which threaten the survival of the brain will be the 'old' brain in a 'new' body—perhaps somebody else's, who has no more use for it, because his brain is defective (or inferior?). If the preferred tendency is towards artificial organs, the last step is to replace everything except the brain by artificial systems—finishing up, perhaps, with the isolated human brain in a cosy bottle, communicating with the outside world via a computer and receiving at will the stimuli corresponding to familiar experience and pleasures—not forgetting orgasms. In hesitant early steps to both such ends (their immediate purposes were much more limited), Robert White and his fellow experimental surgeons of Cleveland succeeded first in keeping the disembodied brains of monkeys and dogs alive for several hours on the laboratory bench; then in transplanting the live brains of dogs into the necks of other dogs, where they were maintained for up to a couple of days by the blood supply of the recipient animal.

One procedure might be to combine brain transplantation with the idea of 'immortality by proxy' wherein, as mentioned earlier, a human adult should be able to produce a much younger twin. Why not keep him anaesthetized while he grows to manhood, then have your brain substituted for his? There will be no difficulty about the graft rejection, because you are, genetically, identical twins.

One snag is that your brain may be getting poorer than it was. The more prosaic science and clinical practice of gerontology is concerned with reducing the deteriorations of old age. Many gerontologists affirm that their aim is not to prolong the span of life—let alone to secure immortality—but merely to keep people alert and healthy until they die at a natural age. Nevertheless, it is hard to believe that success in keeping tissues and organs 'young' will not lead to an extension of life. Like many disorders of the body, ageing seems to be ultimately a matter of chemistry. Tissues harden, for example, because inappropriate molecules accumulate in them. To analyze exactly what happens to us and why, as we grow old, is to go three-quarters of the way towards 'solving' the problem of old age; it will then be for the biochemist to offer corrective potions or injections to keep us young. The 'elixir of life' sought by the alchemists is creeping back on the research agenda. It may, for example, be a compound that mops up destructive 'free radicals' within the cells of the body. If the general tissue deterioration can be kept in check, there seems no reason why men should not live, say, two hundred years. Perhaps our grandchildren will have this option.

Compared with those routes to near-immortality, the programme of the cryo-resurrectionists is bizarre. James Bedford, a psychologist, died in California in 1967 of liver cancer, and became the first man to be frozen with a view to subsequent revival when advances in medical science would make it possible. 'Freeze, Wait, Re-animate' is the slogan of one of the societies promoting the idea. At a cost of $300 a year, Bedford is now stored at the temperature of liquid nitrogen, wrapped in aluminized foil, lying on a plastic bed in an insulated capsule, waiting for some future doctor who can not only revive him but repair his diseased and frozen body—assuming that the caretakers remain diligent until then in topping up his liquid nitrogen. The costs of storing large numbers of people in this way, and of reviving them if the long gamble should pay off, make it a dubious venture for anyone except rich people with a morbid fear of death.

More plausible and perhaps more acceptable is a related idea. If the techniques of freezing without damage can be perfected, living volunteers might be put 'on ice' for revival in a hundred or a thousand years, as 'time-travellers'. They would be able to enjoy (after no subjective lapse of time) life in the future, and explain our curious ways to the historians of that time.

If only a minority of the athanatic technologies summarized here comes to practical fruition in humans—and some of them are mutually contradictory—there will be plenty of moral, legal and political issues to perplex us. How are we to disentangle them?

Notice first the lack of sharp edges to help the legislators who might favour this and disfavour that. There is no boundary between measures accepted in the past and possibilities for the future. The image of a disembodied brain, for example, may disquieten us, yet we strive to keep people alive who are paralyzed from the neck down and whose bodies might be better replaced by pipes and pumps. A clear distinction is also denied us between medical and non-medical uses of the techniques. For these reasons we cannot expect simple yes/no moral answers, nor hope that total prohibition or total permissiveness can cope with athanatics.

Suppose, for example, that law-makers decide that no human baby shall be conceived except by the natural sexual process, or carried till birth except in the uterus of the woman who supplied the egg; that any child produced otherwise shall be deemed doubly illegitimate and the parties responsible for its creation shall suffer severe penalties. Should not present insemination techniques for relieving sterility be exempted? The thought of such an exception illustrates how temporary and circumstantial are our judgements. We happen to have grown used to the idea of artificial insemination in such cases, so we may not only tolerate it but wish to preserve it, rather than pointlessly to deny the satisfactions of parenthood to a minority. In cases where the woman is the sterile partner, implantation of a donated egg merely completes the sexual symmetry, without involving a new principle.

Politically, it may be prudent to mistrust those who think they know how men can be 'improved'. A long-term view—over the decades, let alone the millennia ahead—should make us doubt whether present admiration for efficiency and intelligence, as key virtues in an industrial state, is more than temporary. Yet there will be very crude short-term ambitions, emanating from political leaders, to breed smarter generals or physicists than the presumptive enemy's; to develop powerful

---

V. HALF A DOZEN ATHANATICAL RIDDLES

1. If Dad, a millionaire, had no part in my conception, have I any claim on his estate?
2. If half the organs in my body are replaced by other people's, or animal organs, or factory-made devices, am I still the same person that I think I am?
3. If my brain is stretched by an ingenious biologist, and I invent a whole new branch of mathematics, who takes the credit—he or I?
4. If I am an experimental superman, stacked with non-human gene combinations, have I any rights as a 'human being'—supposing they don't like me and want to kill me?
5. If I could live indefinitely—a couple of centuries anyway—by one trick or another, but the young people won't let me, are they murdering me?
6. Conversely, suppose a cryo-medico freezes me alive, with a view to my revival in the twenty-fifth century. My heart and brain stop, of course. Is he a murderer?

---

infantrymen or astronauts with stretched limbs and high-power hearts; to breed rebelliousness out of the population.

Some kind of equilibrium, well away from extreme attitudes, between those who broadly favour the development of athanatical methods and those who broadly disfavour it, may be the best we can hope for, in the social control of this, the most radical of all technologies. Preventive eugenics, in particular, may be wholly desirable. The natural flow of life, going on irrespective of human will, was interrupted when Onan invented birth control. But confronted now with means not just of switching life off and on but of altering human life in far-reaching ways, we may have to re-affirm satisfaction with the qualities of natural men—not despite the unpredictable and awkward variations but, in large part, because of them.

BETTER THAN CURE

Heart searching about heart transplants in the late 1960s showed that, even if many athanatical possibilities were still remote, plenty was already coming out of the research pipeline, with moral and political consequences. The French cabinet was obliged to give a ruling on a new definition of death (when the brain stops, not the heart) so that human-to-human transplants could proceed. But that, too, was technically dubious.

Questions of professional ethics, as intricate as the scientific procedures that give rise to them, litter the medical landscape —and not only from polite forms of cannibalism like organ transplants. Decisions about when a patient is to be allowed to die, or when a man is to be rated low in priority for expensive and time-consuming therapy and thereby sentenced to death, have to be faced every day by hospital doctors. Research has put at the disposal of physicians so many techniques that increasingly lack of material and manpower, rather than lack of knowledge or skill, limits what can be done to save life, relieve suffering or avoid unnecessary incapacitation. 'The doctor's dilemma' is not a new phenomenon; it will never be eliminated at the front line, in the wards, as long as men fall sick and physicians are overworked. As it is a big and complex subject in its own right I propose to leave it aside.

Issues remain, however, that concern the strategy of health rather than the tactics of therapy and lifesaving. They will figure prominently in the politics of the 1970s.

Some of the new techniques and associated intensive care are far too expensive for ordinary people to be able to afford, without mortgaging the rest of their lives, and probably beyond private insurance cover if their application becomes widespread. In those countries that do not already have comprehensive free public health schemes, a sense of growing injustice in matters of life and death will over ride the political and professional opposition to 'socialized medicine' in the USA and elsewhere. It was significant that the first heart transplant in Britain was done 'under the National Health Service'.

A special effort to remove the element of the market-place from medicine may be necessary to prevent a trade in vital organs and tissues. Just as, in the past, hard-up young women would sell their natural teeth for dentures, so many people may be bribed to part with a kidney or an eye. And just as Burke and Hare committed murder to supply the Edinburgh dissecting theatre with cadavers, so others may find ways of providing valuable organs too conveniently. This problem will be greatly eased if there is justification for present hopes of medical researchers that, using the anti-lymphocyte serum, it will be possible to transplant organs from animals to humans.

Even where health services are provided from public funds, some arbitrary limit will have to be set to the proportion of the nation's wealth allocated to them. It may be considerably higher than at present, but it will not be enough to do all the things that might be done for all the patients. Some budget limitation may indeed be desirable to discourage endless tinkering with the human organism with drug and knife, in a mood of communal hypochondria, and to maintain a sense of perspective in the medical profession.

The only real hope for reconciling the potentialities of medical science with budgetary realities is to reduce the number of serious hospital cases by a comprehensive programme of preventive medicine. The professional energy of the physician is directed towards treatment rather than prevention of disease. Dramatically successful exceptions, in the vaccination and screening programmes that have virtually eliminated infectious diseases like smallpox, tuberculosis and polio from many countries, merely highlight the rule that most of the rest of medical practice simply waits for cases to turn up.

The heart defects and cancers that are now principal causes of death in rich countries are much more susceptible to treatment if spotted when still incipient. Physical and biochemical instruments have already been developed which would permit intermittent monitoring of the whole population of apparently healthy people, without inconvenience. The costs of setting up a full-scale programme might be very high but, as therapy becomes more complex, the justification will become overwhelming.

Since the nineteenth century, increasing attention and expenditure have been devoted to public health services and preventive medicine. By and large, those countries favouring rugged private enterprise and opposing massive public investment in social services have lagged in preventive medicine. If the basic pattern is one of the patient going to the doctor when he feels ill, receiving treatment and paying cash for it, the medical profession has a vested interest in disease. Conversely, whatever hard things can be said about the USSR, its health services are magnificent.

In the near future, preventive medicine will become more effective, more expensive and more revolutionary, and hence more controversial. It will involve large-scale redeployment of medical resources, both financial and human. On the technical side two possibilities may be mentioned. The first is the use of biochemical diagnosis, which involves the sampling of blood, saliva, breath or urine and the study of metabolism of the body as disclosed by the chemical constituents of the fluid. Universal tests of this kind would require elaborate equipment provided on a large scale, though not necessarily a substantial body of physicians and medical technicians, if the analyses can be automated. They would enable detailed computer records to be kept of the state of the body in each citizen, and any incipient diseases would be detectable. Such a technique, reinforced with X-rays and heart checks, could greatly reduce the incidence of disease and, providing early detection, would greatly simplify the cure. Another potentially expensive but effective technique would use 'socially acceptable monitoring instruments', electronic recorders which would be worn by individuals for, say, a week each year and would show precisely how the heart and other organs performed during normal working life. These, too, would help to detect major diseases well in advance of the appearance of obvious symptoms.

The provision of these and other techniques on the necessary scale is a matter of political priorities and deployment of manpower. Politically much more sensitive, perhaps, in the long run will be the growing understanding of the relationship between health and the general social environment. The connection between dirt and infection, between malnutrition and vulnerability to disease, the diseases of malnutrition itself, and the high incidence of accidents associated with road transport and particular industries—all these are well recognized. Action based on this knowledge has helped to make life more tolerable and healthier in all countries and has particularly encouraged the development of welfare states.

The United States was the first country to insist on the printing of cigarette packets with notices that cigarettes were injurious to health. In Britain, where the connection between cigarette smoking and lung cancer was first formally established by medical researchers, the government took a more timid view and went no further than banning the advertising of cigarettes on television and mounting its own modest anti-smoking advertising campaign. The tobacco industry is a major enterprise and source of public revenue in most countries; decisive action warranted by the medical evidence, such as the total banning of the manufacture and sale of cigarettes, would have important economic effects and would, of course, arouse the great displeasure of the large number of cigarette smokers. But this is precisely the kind of measure to be expected if preventive medicine were to be taken seriously.

Similarly, rigorous measures against road accidents would entail, for example, a maximum speed limit of sixty miles an hour on all roads, banning from driving of people convicted of motoring offences, exclusion from driving of people under, say, twenty-five and over sixty-five, and other unpopular measures including abolition of motor cycles, prohibition of the manufacture of cars with speeds greatly exceeding the maximum speed limit and the redesign of motor vehicles to give their occupants a better chance of survival in the event of an accident.

But preventive medicine can go farther than that, if and when its exponents acquire power and opportunity. All the conditions and stresses of modern life, as determined by the physical and chemical environment, by transport systems, by methods of organization and work, even by sports and entertainments, could become subject to the scrutiny of medical authorities. Issues of birth control, city planning, treatment of children, and so on are at root interpretable as health matters. So is the negative eugenics, discussed earlier.

Preventive medicine may be a powerful aid in exerting social control over the uses of science. Environmental elements such as food, radiation, drugs, pollution and conditions of work have important and long-lasting effects on people, but they tend to be subtle and unobvious effects. They can only be revealed by very careful study of human beings and their health, or, in some cases, by animal experiments. For example, pharmacologists subject new drugs to extended animal tests, looking for marked effects of the drugs. To find subtle influences may involve the use of much larger numbers of animals over long periods of time —'megabiology', the kind of work done at the Oak Ridge laboratory on the long-term effects of radiation, in which 200,000 mice were kept under observation throughout their lives. Big, specially built laboratories, using doctors and scientists from a wide range of disciplines, will be needed for such work.

Drastic interference with the lives of individuals will occur in the name of preventive medicine. Major political battles in the decades immediately ahead will centre on the issues of how free a man is to endanger his own health, and what responsibilities public authorities have to stop him. Arguments about social priorities will determine the extent to which demands of health will be allowed to dictate, for example, town planning, managerial operations, educational techniques and so on. Carried to extremes, preventive medicine could itself become tyrannical, giving authorities ever widening powers to monitor personal behaviour and tell people how they must live.

CHEMISTRY VERSUS INDIVIDUALITY

MAO may be a name for the politically alert to conjure with when the Chinese chairman is forgotten. It stands for monoamine oxidase, a type of natural catalyst in the human brain. It is one of the enzymes that regulate a man's state of mind by keeping a check on the supply of 'transmitter substances' whereby brain cells stimulate one another. The action of reserpine, a tranquillizer, is evidently to release transmitter substances to attack by MAO. As pharmacologists discover how drugs exploit or inhibit the activities of MAO and other enzymes in the brain, so they can make the control of mental conditions by drugs much more precise.

There is probably no way in which technology can touch an individual more intimately than by altering his mental condition by psychic drugs. The power to do so grows steadily. Conventionally accepted drugs—alcohol, nicotine, caffeine—serve individual and social purposes, but their effects are mild and imprecise compared with what will be possible in the future. Thoughts of MAO may serve to raise a general question in a particular way: is individuality going to disappear?

Mass production of education, of jobs, of entertainment and of consumer goods makes lives in a rich country increasingly similar. Yet the individual's response to his environment, his moods and his thoughts depend on subtle variations in MAO activity and the other processes in his brain. No two individuals are likely to respond in exactly the same way. But bring MAO and the rest under sweeping voluntary control by drugs taken by large numbers of people, and it will be like turning the wilderness of human thought and emotion into a garden. Here are roses of tranquility, here the narcissi of purposeful concentration, there the hemlock of sedation, just as the patient ordered. Whether that is a good idea or not depends upon psychological and political attitudes. If the individual is primarily a member of a superhuman society, then he will doubtless be a more congenial member if he controls his mood to suit his colleagues and neighbours. If society is primarily a collection of individuals, loved and respected for their individuality, such control is apparently noxious except for the treatment of grossly diseased minds.

The chief branch of research that explicitly accords respect to the individual is psychology, in its inquiries into the mental needs and drives of the individual brain. At present it is an underdeveloped science or art, with practitioners having to sift eclectically through rival theories and questionable data. Advances in brain physiology should provide a firmer basis for exploring human thoughts, hopes and anxieties and testifying to the uniqueness of the brain, as the geneticists and immunologists have proclaimed the uniqueness of the genes. But, anyone holding up psychology in defence of the individual must not forget that, misused, it can break an individual. If it is a matter of teaching, persuading or forcing people to conform to conditioned patterns of behaviour, psychologists have plenty of tricks already. In future, drugs will make control easier.

The human brain is about the most complex system known to modern science, and the combined efforts of biologists, biochemists, doctors, engineers and mathematicians are only slowly disentangling the complex processes involved in its daily workings. But already it looks probable that much serious mental disease will be susceptible to chemical treatment. Long-term memory turns out to be a matter of chemistry and in the mid-1960s there were claims that particular materials enhanced the memory, for example in aiding old people with failing memory. Other strange and controversial experiments suggested that it might be possible to switch learned information between animals, by transferring material from the brain of one into the blood stream of another. (Hence the suggestion that professors should be minced up and fed to their students.)

Psychic drugs will, if desired, provide the short cuts to other sensations. 'Trips' by drug may be an escape to a much-wanted inner sanctuary of privacy. But drugs will also enable users to control their moods and responses in a way deemed socially appropriate by others. One can, for example, envisage widespread tribe-like rituals in which everyone takes his pill and all share a common feeling. The basis for such expectations is not so much the potent effects of accidentally discovered drugs like LSD as the current systematic elucidation of drug action and brain mechanisms—involving MAO and the rest.

Few decisions about the uses of technology are likely to have as great an effect on personal life in the future as the choices about the acceptability and role of these psychic drugs. Leaving aside directly military or tyrannical uses of drugs taken involuntarily, there are three main possibilities:

1. Drug-taking for social reasons—typically the future analogues of caffeine, alcohol, nicotine, tranquilizers or pep pills.
2. Drug-taking for personal pleasure of 'escape'—typically, the future analogues of pot or LSD.
3. Rejection of drugs except for serious medical purposes.

It remains an open question how far the decisions will be left to individuals or brought under legislation, and whether different patterns will co-exist in one community.

The first possibility is likely to commend itself to those who take a community-oriented view of life. The more individualistic will be split into those who want the alleged additional 'dimensions' of inner life from drugs, and those who see drugs as a threat to their mental integrity. The latter may accept the paradox that one must circumscribe an individual's right to use drugs, in order to protect his freedom as an individual.

There will be no easy ruling to make, as there is in the case of prohibiting 'hard' narcotics. No amount of argument, unless it be about ill-effects or side effects, will arrive at a 'right' answer about psychic drugs. It depends upon one's view of what life is all about. But argument and practical decision there must be, because the so-called 'soft' drugs are already with us, and more are on the way.

CHAPTER 13

Pigmented Poor

GENERAL WINGATE ROAD in Addis Ababa dips down into a valley just in front of the modern building that houses the parliament of a feudal country. Topography and eucalyptus combine to screen the valley bottom from the distant eye, but a short walk brings you into a broth of tumbledown shanties; not the worst conditions of human life in this naturally beautiful but socially squalid country, but nauseating enough and close enough to the shining palaces of government and commerce to induce a sense of outrage. Drivers in automobiles worth a lifetime's earnings for the average Ethiopian take this short cut to the Piazza, hooting absent-mindedly at the woman and the naked baby girls of General Wingate Road. In the scientific age, those children face an ignorant life of hard labour and scant reward.

Illiteracy rates in Ethiopia are among the highest in the world and the few students absorb the standards of their well-to-do fathers. One science freshman at the Haile Selassie I University, from the northern Tigres tribe, remarked to me of the Ethiopian poor: 'They have no education so they do not want anything more.' A professor of public health took his students to visit the foci of disease in the Addis slums and the would-be doctors protested: 'Why do you bring us here? We shall never have to treat people like these.'

But education evokes other sensibilities, and in April 1967 students at the Haile Selassie I University and the Teacher Training Institute demonstrated against a new law restricting demonstrations. Police armed with guns, tear gas and batons attacked the students in force, brutally beating even those who were not demonstrating and throwing hundreds into jail. In an orgy of violence they smashed windows and equipment in the colleges and hostels and looted property of the students and staff. According to a report by staff of the Teacher Training Institute: 'The hatred of students and teachers indicated by the actions, abuse, facial expressions and remarks of the police suggests that the police had undergone considerable anti-intellectual indoctrination.' Representatives of the landowning ruling class on the university's board of governors wanted to close the university for good.

Yet to wish for revolution in this oppressed, corrupt and metastable country, propped up from Washington, may be to countenance regional fission and bloody war between the several tribes, Semitic and Negro, or between Coptic Christians and Moslems. What, then, do you do about a country like Ethiopia? There is no alternative to massive aid, irrespective of politics. The present influx of foreign doctors, teachers and industrialists, some of whom make heroic efforts, offers only palliatives, and superficial achievement in showpiece hospitals and factories. Meanwhile, the population grows. In rural areas, girls may still be married at the age of five or six, and have their first babies at eleven.

What had taken me to Addis Ababa was an international scientific conference, concerned with the application of microbiology to the benefit of the poor countries of the world. It was very constructive, most memorably in discussions of the cultivation of micro-organisms as food. But the exploitation of such ideas to make a significantly better life for the African poor seemed a formidable task. I felt a similar despair when hurrying in the company of an Indian scientist from one modern laboratory to another. We had to take a train from Calcutta one evening and, as we walked up the platform, he talked without pause of plans for the Indian electronics industry. He seemed oblivious of the fact that we were having to step cautiously over the bodies of the many men, women and children who were sleeping beside their belongings on the platform, because that was their home.

In these examples, I do not mean to deride the technological ambitions; on the contrary, one of the troubles is that there are not enough of them. Given the best will in the world—which we are not—the remedies for poverty would be hard enough to contrive. As things stand, apathy, geopolitical factors and defects in the world's monetary and trading arrangements make matters worse. The rapid growth of population in the countries that can least afford sustenance for the newcomers is alarming. But the most dismal aspects of all are the racism that underlies the division of the world into rich and poor nations, and the negative part played by research and technology.

No caricature is involved in describing modern science as a European invention which enabled the white nations to achieve military, economic and cultural domination over the rest of the world, and to make themselves prosperous while leaving the natives of the poor countries to progress very much more slowly. No injustice is done, to say that most research workers and technologists have unthinkingly connived in these uses of science which are, at bottom, racist. Declarations about using science to feed the world's hungry have not stopped the prosperity gap growing wider; nor can they alter the fact that the intellectual interests of the great majority of research workers are far removed from any such programme, or that the preoccupation of technologists is with machines that enrich the rich.

Domestically, too, the same sort of inequity occurs. The plight of the urban negroes of the northern United States was in large part a direct consequence of technology. Agricultural technology so greatly reduced the requirement for farm labour in the southern states (where, in 1940, three-quarters of the negroes lived) that half the negroes in large northern cities were immigrants, chiefly from the south. They no sooner became established there, in unskilled and semi-skilled jobs, than the technological advances in industry both reduced the number of such jobs and encouraged the creation of new factories away from the city centres, out of reach of the urban ghettoes. In 1967, the unemployment rates for negro teenagers were over thirty per cent, twice that of whites for the same age group.

There is probably more productive research of intended value to the poor, coloured countries going on in the rich, white countries than in the poor countries themselves, but it is a small part of the total effort. As individuals, distinguished scientists of the rich countries have done more than anyone to sound the alarm about growing populations and the threat of intractable hunger and poverty in half the world. John Boyd Orr, Julian Huxley and Patrick Blackett were among the early prophets of world development. As individuals, too, many young research workers and technologists would be happy to do more to remedy the situation if that were where openings existed for them to work. But a moral and political failure on the part of the rich nations to respond adequately to the crisis in world development leaves most of educated human creativeness dealing with mini-problems of the white rich.

It is worse than unhelpful. In several major ways the educational, research and development policies of the rich countries have operated directly against the interests of the poor. Four effects are worth discussing:

the brain drain from the poor to the rich countries;
the irrelevancy of much research even in the poor countries;
the development of synthetic or substitute materials that replace imports from the poor countries;
the equivocal attitude to birth control.

Richard Titmuss (1967) estimated that the foreign aid received by the United States in the form of scientifically trained manpower (1949–57) saved that country $4000 million; he accused the rich countries of deliberately enticing qualified manpower from poorer countries instead of training more of their own youngsters. The very countries that will gladly recruit an Indian Ph.D., an African B.Eng. or a West Indian M.D. will mercilessly turn away a carpenter or labourer of the same origins.

The fashions in science are set exclusively in the rich countries and reflect the local intellectual and social interests. Research workers in the poor countries, anxious to make their reputation on the world stage, are strongly tempted to follow these fashions; international programmes assist them in doing so. Research which bears most directly on the social and economic needs of the poor countries tends to look unfashionable if not pedestrian. It would be unfair and counter-productive to discourage the poor countries from trying to shine in branches regarded by the scientific world as 'advanced'. Only the research workers in the advanced countries can mend matters, by bending their minds to the global problems. To put it crudely, a major discovery about the dynamics of the monsoon should be as eligible for a Nobel prize as new results about the dynamics of the mu-meson; or immunology of malaria as eligible as immunology of cancer.

The development of synthetic fibres and synthetic rubber threatens the traditional cotton, silk, jute, flax and natural rubber exports of the poor countries, and the rural economies that produce them. In a more equitable world, not divided into rich and poor, the displaced producers would benefit in the general economic gain from the better or cheaper synthetic product. But they do not. Again, developments in nuclear power and restrictions on the use of polluting oil will favour countries producing nuclear fuels at the expense of the present oil producers. The economic situation is so complicated that it is hard to imagine a coherent policy developing for control of technology in such respects. There may be occasional blatant cases deserving some special intervention. For example, if, as is likely, food technologists develop synthetic coffee, cocoa and tea, as good as the natural products and perhaps cheaper, the effect on the economies of several poor countries could be so catastrophic that some international agreement to postpone the marketing of these materials would have to be sought.

For two decades after 1945, while demographers were pointing out that for every two human beings who died three were being born, religious opposition from the Catholics and simple prudery from the Protestants made it very difficult for investigators of human reproductive physiology in the rich countries to obtain government sponsorship for research on birth control. They did a certain amount under the guise of other research, but the Pill and the Loop, the two major innovations, were developed and proved with private sources of funds. Internationally, too, birth control was an unmentionable subject as late as 1963 when, at a big UN conference on the application of science and technology for the benefit of the 'less-developed' areas, papers on birth control were excluded. In the mid-1960s, the attitudes improved markedly but, in the meantime, the most obvious line of research relevant to the problems of the poor has been wilfully impeded by the rich.

The population explosion is not a matter of dead statistics; it is a lot of live and hungry children. What is more, it is a lot of young girls, like those in General Wingate Road who, in the 1980s, will be producing yet more children. Between 1968 and 1983 the number of women aged between twenty and twenty-nine—at the most fertile period of their lives—will approximately double.

Parents in the poor countries have large numbers of children for the good biological reason that so many of them die young. We are in a transitional phase, when the death rate has been cut, but neither sufficiently nor for long enough to convince parents of the need to cut their family size. Maximum visible effort in keeping young people alive may, paradoxically, be the best way to ease the growth in population.

An international treaty may be needed, whereby governments will agree on target ceilings for their respective populations. Social inventiveness may be wanted, too, like the Indian policy of giving transistor radios to men who volunteer for sterilization. Kingsley Davis (1967), urban research worker at Berkeley, is much sterner: he wants to make the most of factors (like housing shortages, military service and obligatory work for women) that help to postpone marriage and reduce child bearing. He would have the sixty-two governments which give cash payments to parents for their children reverse that policy and levy a child tax instead. Yet perhaps such high-level policy is irrelevant, and the crucial step is to reach the mothers through health services and approach family planning as a matter of good medicine.

SCIENCE DOES HELP

The biggest turnabout in the post-war struggle for world development was marked, in 1968, by the setting up of a working party within the Food and Agricultural Organization of the UN to study the implications of high-yielding crop varieties. The chief motive was concern about the impact in India and Pakistan of high-yielding grain of such quality that it promised a rural revolution. These countries which, in the mid-1960s, had been almost written off as far as any long-term capacity to feed themselves might go, were introducing wheat and rice seeds that suddenly made it possible. Years of diligent work by planners and agricultural specialists had been frustrated only by lack of such seeds; provision of fertilizer, water, pest control and rural education wanted only this ingredient for success on a startling scale. Then the planners in the Indian subcontinent and FAO had quickly to bend their minds to transport of local surpluses, maintenance of prices and rural developments that would consolidate the success. The problem of the empty granaries of Asia seemed possible of solution—at least for long enough to give population-limiting policies time to take effect.

It was a triumph for scientific plant breeders, in American-sponsored work that began with the development of high-yielding hybrid maize and went on with spectacular developments with wheat in Mexico and rice in the Philippines. There could be satisfaction that, through the Rockefeller and Ford Foundations, profits from oil and automobile manufacture were diverted to alter the course of agronomic history at its most critical moment.

Other dramatic news came from Cuba. Fidel Castro's regime brought out from Scotland the 'barley beef' pioneer, T. R. Preston, to direct an Institute of Animal Science. Two problems coincided: the dependence of Cuban economy on sugar and the shortage, throughout the poor countries, of animal protein. Very quickly, Preston and his colleagues developed a feed for cattle based on molasses and urea—the simplest form of organic nitrogen for protein formation. The yields of beef on this diet, and the high yields attained in sugar growing, made it possible in principle to triple the production of meat from a given area of farmland.

Agitation about the world food shortage, among a relatively small number of scientists, was sufficient also to open up quite novel sources of food. For example, a British biologist, N. W. Pirie, beginning in the 1940s, demonstrated that edible protein could be extracted from ordinary leaves that would otherwise go to waste. And at the Addis Ababa conference of microbiologists, mentioned earlier, French and British experimenters reported the first successful trials of feeding livestock on a diet of yeast produced from petroleum. There was excitement, too, about the recent discovery of a tribe in the Sahara who harvested mats of an alga called *Spirulina*, and ate it; scientists on the French Riviera were cultivating protein-rich *Spirulina* in tanks. Some of the biologists went out to Lake Aranguadi, not far from Addis, covered by the same alga as if it were green paint, and Carl-Göran Hedén of Stockholm suggested that it was perhaps the most efficient protein-producing organism in the world.

Given such developments and discoveries, on the basis of work by a few dedicated men, what would not be possible if more of the world's research and technology were devoted to the needs of the poor? At the same time, it would be disastrous to suppose that science can do the trick without any other exertion on the part of the rich. All that relevant advances in knowledge and skills can offer is hope for the solution of otherwise intractable problems. Their fruition depends upon abundant finance for practical application, and even then they do not begin, by themselves, to accomplish the necessary reconstruction of antique cultures.

In that connection, indigenous science, even if less spectacular in its results, may be more important in the long run than contributions of rich-man's science. Consider one of the most depressed groups among the whole of humanity—the Untouchables of India. To them falls the traditional work of flaying the sacred cows that have died of old age and, amidst stench and stain on the outskirts of remote villages, they make leather. Yelavarty Nayudamma, chemist and director of India's Central Leather Research Institute, has therefore one of the most daunting tasks facing any man of science, in his attempt to modernize this traditional but economically important industry. But, from his laboratory in Madras, he embarked on bold policies. He sought to create a common culture of researchers and tanners. He himself would go off to the remote villages and tanneries, each month to a different part of India, to teach but also to learn about local problems. For a civil servant, Nayudamma's methods were unorthodox. He would find the most receptive tanner in an area and would give technologies that would make his competitors jealous, and therefore eager to imitate. Or he would take over an unsuccessful firm, make it profitable and give it back. He found ways of explaining technologies to tanners who were illiterate, never mind scientifically trained. All this he did against a background of pioneering research in his own institute.

Equally striking, in a quite different way, was a flowering of advanced science in a poor country. In Mexico after the second World War, a small company, Syntex, was created to exploit a material called diosgenin occurring in certain Mexican plants. Russell Marker of Penn State had found it could be transformed into medically useful steroid compounds, including hormones. In 1949, when cortisone became a medical sensation, the activities of Syntex expanded and the company created a research department. At that time no Mexican university offered post-graduate training in chemistry. Syntex did its own post-graduate training and also subsidized the institute of chemistry at the national university. For work that required chemists of doctoral level, such people were brought from abroad. By 1959, according to Carl Djerassi (1967), more scientific publications in steroid chemistry had emanated from Syntex in Mexico than from any other academic or industrial organization in the world: 'Mexico—a country in which no basic chemical research has been performed previously—had become one of the world centres in one specialized branch of chemistry.' The work included the development of oral contraceptives. Moreover, by 1960, well over fifty per cent of the world's supply of steroid hormones originated from that country. Within Mexico, clinical research, botany and veterinary research had a great fillip.

Mexico is one of the few poor countries that show real signs of becoming rich. For others, the dead weight of poverty will take a Herculean effort to lift—an effort, involving the whole species for many decades to come, for which there seems little inclination at the time of writing.

THE RACISTS

Winds high in the Earth's atmosphere are responsible for the differences in human skin colour that make men mad. About thirty miles overhead, where the air is very thin, a good deal of ozone is formed from oxygen by the action of sunlight. Without this ozone layer life on the land would be impossible; indeed plants and animals originated and evolved only in water until enough oxygen existed to draw the veil of ozone across the sky. Ozone absorbs most of the ultraviolet rays from the sun—rays that would otherwise be fatal to all living organisms. Even so, it does not absorb them all, and human beings, evolving as they did in a sunny region, had a pigment, melanin, in their skin that protected the underlying tissues from the ultraviolet.

Winds in the ozone layer blow from the equator towards each of the Earth's poles. As a result the ozone is depleted over the tropics and heaped up over higher latitudes. In the tropics, the ultraviolet is most intense and those members of the human species with the darkest skins had the best chance of survival in those regions. In higher latitudes, another physico-chemical factor came into play. A little ultraviolet is good for you, because it stimulates the production of vitamin D in the body; without that vitamin children suffer softening of the bones known as rickets. In the higher latitudes, with relatively little sunshine and a thicker ozone layer as well, heavily pigmented skin was therefore a liability and paler-skinned people prospered.

Thus a single species adapted itself to different physical conditions in different places. All that is in the past tense, because clothing, hats, sunglasses and cod liver oil dispose of the selective pressures. Apart from collateral variations in stature, girth, eye colour, hair and so on, that is about all there is to it—except for the political capital men have made out of the conspicuous differences between them.

Forms of racism are probably as old as tribal warfare and to this day there is a close affinity between racism, nationalism and war. The really dramatic event, however, which established the present peck order was the colonial expansion of the Europeans. Among birds of a hierarchical species, social ordering may be established by the dominant male pecking at the other birds and driving them away; subordinates in turn peck at their subordinates. A similar peck order exists, *de facto* though not *de jure*, between the so-called races of men. The concept of race is shaky from a biological viewpoint, because of the pronounced admixture of genes, but it would be silly to pretend one does not know what it means politically. At the top of the present human peck order are the fair northerners, followed by other paleskins, then by the yellow-skinned, the brown-skinned, and darkening shades of black, with the Australian aborigines and the African pygmies at the bottom of the list, with no one left for them to peck at, supposing they wanted to do so.

Some such ordering may have roots deep in prehistory and in the fertile imaginations of men. But it was when Europeans, equipped with Chinese technology like the compass and gunpowder, set out to annex other parts of the world, killing or enslaving the native populations, that the pecking began in earnest. The associated rise of modern science confirmed and strengthened the ascendancy of the Europeans around the planet. Racists use the argument of demonstrable scientific and technological superiority of the pale-skinned conquerors as evidence of innate superiority. The fact is that modern science had to be invented *somewhere*; until it was, people of the relatively prosperous regions differed remarkably little in their attainments, if one allows for cultural crescendos and diminuendos.

Till the European phenomenon, the Asians had perhaps done most, but attainments in Africa and pre-Columbian America were far from negligible.

The Indians' caste system and disdain for Africans, the mutual hatred among African tribes, and the many other irrational subtleties of human chromatography, show that racism is no monopoly of the whites. Nor is skin colour the only stimulus for prejudice. Hitler's extermination of the pale-skinned Jews with poison gas was a logical outcome of hatred. The pre-judgement of a man's worth according to his ethnic antecedents is morally and politically the worst of all blights on the face of twentieth-century civilization.

Psychology, anthropology and social research have much to contribute to eradication of personal prejudice and discrimination in social systems, but they can, at best, take effect slowly, making people colour-blind over a minimum period of one or two human generations. The down-trodden populations will not be so patient, and their revolts, actual or incipient, will tend to harden prejudice. However much those who have no conscious racial prejudice may wish to ignore a stupid problem and hope it will disappear, it will not do so. The riots in American cities, the growth of racism in Britain, the tension in Southern Africa and war in other places suggest that the problem must resolve itself in some radical way. The possibilities include the following five:

Major war or genocide. A chronic world war of the races, aggravated by economic difference between the rich, white north and the poor, coloured south is unhappily one of the more plausible predictions for the next few decades. The whites have a near-monopoly of the most powerful weapons at present, but biological weapons, to which the underprivileged countries will have easier access, may seem well adapted for race war. In particular, they could exploit genetic and immunological differences between nations. Perhaps the most lamentably ingenious perversion of science one can imagine is a man-made disease agent that is inactivated (or activated) by skin pigment. Even if one ethnic group were somehow eliminated, it would remove only one step in a long peck order.

Resegregation. There is no convincing precedent for a well-integrated multi-racial society so that anyone pessimistic about the possibility of inventing one might favour a voluntary kind of *Apartheid*; for example, the creation of 'all-black' cities or states within white majority countries. Only if full economic equality and interchange were assured, which would be difficult, could such a policy begin to interest the underprivileged. On the other hand, wars against foreign intervention in the poor countries, and policies of 'Africanization', tend towards a kind of self-segregation. It is not very compatible with the growth of international technological enterprise and co-operation.

Confirmation of ethnic rank. Hypothetically research, more probably phoney than genuine, might establish evidence of superiority and inferiority sufficiently convincing for people to accept a lowly status on account of their 'race'. The existing evidence is that any differences are far too small and subtle for that. Scientifically the idea is quite far-fetched, but socially it is not, because even among some dark-skinned people there is often a deep-rooted idea that lightness of skin is a virtue or a sign of superiority. Educational psychologists find that children thought by their teachers to be untalented act accordingly, to please.

Eradication of ethnic differences. Voluntary intermarriage, if neither promoted nor discouraged, will inexorably, but very slowly indeed, produce a more uniform species. A thousand years hence the effects of that still incomplete process may be regretted because they are boring. In any case they are almost irrelevant to the present issues. What unhappily seems a likely and quickly applicable use of impending biological knowledge will be for the chemical or genetic intervention in human beings for 'cosmetic' purposes. For example, it is more likely than not that some safe way of blocking the synthesis of skin pigment will be discovered, perhaps using antagonists of the enzyme tyrosinase. Some black women may choose to have white babies, by embryonic manipulation, to give them a better chance in life. The results of such interventions may often be pathetic or grotesque and are unlikely to help resolve the general problem.

Political and economic equality. This is the only route promising reasonably quick abatement of racial strife. It will be extremely expensive to implement. Realistically, it can operate as a matter of policy and law without prior eradicaton of prejudice, and it is compatible with either integration or re-segregation. But the problem is a dual one: of equality within each country and between countries of different dominant skin colours. The widening prosperity gap between rich and poor nations can only tend to complicate domestic problems. There are various, though secondary, technological connotations of such a policy. The international aspects are discussed in the next section. In domestic policy, the quest for equality will be closely coupled with the technologies of cities and of education, and also with changes in the products, methods and location of industry, which tend to hit hardest the already underprivileged.

WELFARE WORLD

Pope Paul VI (1967) called development 'the new name for peace'. But, by then, 'aid to the underdeveloped countries' (euphemistically the 'developing' or 'less-developed' countries) had become a tired, deadening cliché that few in the rich countries took very seriously. The first UN Development Decade (the 1960s) fizzled and there was widespread disillusionment among politicians in donor countries because it had failed to produce magical results.

In the late 1960s, the idea that the rich countries would share their wealth with the poor countries on any substantial scale—that is to say, going far beyond the prevailing levels of aid—seemed political moonshine. The dollar and the pound were shaky, and rich countries like the USA and Britain were trimming their aid budgets without any noticeable political controversy. There was a lapse of a generation between fairly wide acceptance, in the rich countries, of the principle of the Welfare State and the first serious, though speculative, attention to the principle of the Welfare World. In the 1970s, after shock of famine or disorder, and with the threat of worse to come as those huge cohorts of girls reach child-bearing age, international and internal political pressure on governments of the rich countries may take effect.

According to an early prophet of the Welfare World, the Swedish economist Gunnar Myrdal (1958), it would not be done voluntarily. He regretted the narrow nationalism of the underdeveloped countries at a time when they should be joining forces to raise their demands and forcefully press them home. Myrdal foresaw regional planning and economic policymaking for the poor countries, and the development of closer commercial relations between them. He castigated the rich countries for their willingness to provide capital aid and technical assistance while refusing to make those adjustments to their economic policies and their ways of doing business which would count for much more—in particular the need to open their markets to the exports of the underdeveloped countries. The very liberalism of the nationalistic Welfare States of the rich Western countries, which gave scope to special groups with commercial interests in influencing national policy, made it difficult to improve the situation. The rich nations had to purge from their minds what Myrdal described as 'the untenable notion . . . that people in the poor countries should continue to be prepared not to think, desire, dislike and act in the same way as they themselves do as a matter of course'.

Harrison Brown (1967), geochemist and foreign secretary of the US National Academy of Sciences, declared that massive capital and technical assistance for the poor countries would be necessary for a hundred years, if the gap in standards of living was to be closed. He wanted the rich nations of East and West to create, under the United Nations, a fund of capital for development, disbursing $15,000 to 20,000 million a year. His idea for encouraging thrift and discouraging arms races in the poor countries themselves was to make the maximum allocation from the fund equal to a nation's domestic savings minus its expenditure on armaments.

Although the sum suggested by Brown is massive by any standards, so is the world population. Divided among the poorest half of the world it amounts to only about five dollars per head per year—the equivalent of one modest meal in an American restaurant. In view of the growing affluence of the rich countries and their massive indulgence in military and frivolous extravagances, $15,000–20,000 million a year must be taken as a minimum figure if it is supposed to represent a serious effort to redistribute the world's wealth. On the other hand, properly invested, it could work great changes, in self-multiplying ways. And other measures could greatly add to the effective contribution of the rich countries, to make this sum, in effect, five times greater.

The monetary, economic and planning aspects of such a programme, including the measures necessary to make it possible for wealth to be exported by the rich countries and productively absorbed by the poor countries, will set limits on expenditure beyond which aid will be of no avail. Here I shall deal principally with implications for research and technology in the rich countries, given a serious Welfare World policy; also with what can already be done, in anticipation of such a policy, to begin switching scientific effort in that direction.

Although primarily concerned with 'righting wrongs', in the attack on needless hunger, disease and squalor, a major effort in world development will involve much new creative activity. The brainpower of half the world is to be liberated from ignorance and preoccupation with subsistence. New styles of scholarship, art and manufactured goods, perhaps new styles and topics of scientific research, reflecting non-European traditions, can be anticipated. Hundreds of new cities and industrial complexes will be created where now there is desert, tropical forest, or weary farmland. The way of life of half the world has to change abruptly but humanely. There is as much work to be done in the next fifty years as in the whole history and prehistory of mankind. Either it will become the chief preoccupation of men and their governments or, neglected, it will make trouble that will be the chief preoccupation. The poor will not vanish.

The strategy for a Welfare World will require technological

change in the poor countries; the most ingenious and determined effort to develop agriculture and other elementary uses of natural resources cannot raise the general level of prosperity very high. Agriculture requires special attention, as the present economic base of the poor countries as well as the source of desperately needed food. At the same time, industrial development of various kinds must proceed, not necessarily repeating the successive phases of industrialization of the already rich nation but certainly taking over manufacture of many products from them. Governments that are revolutionary, in spirit if not by accession, will be needed to lead the poor nations to new ways of life.

A policy of self-denial in manufacture by the rich countries, to make way for imports of goods from the newly industrializing nations, may be the principal supplementary act of welfare policy, beyond the capital aid fund. As a result, the rich countries may cut back such industries as motor vehicles, refining of imported oil, and domestic electronic and electrical equipment; on the other hand, their imports of food may tend to diminish and exports to increase. They will be exporting capital items like production plants, computers, aircraft and so on, embodying advanced technology needed by the poor countries; also designs, technical know-how, professional skills and research services, and information in the form of books, periodicals and electronic programmes.

The goal of the Welfare World, if intended seriously enough to entail some material sacrifice by the rich countries, will reduce their capacity to pursue other technological goals at home. If so the rate of technical change will ease in those countries, even if the technologists are fully occupied with current problems of the world as a whole. The chief personal burden will fall on the professional and skilled workers in the rich countries; a cut-back in 'easy' manufactures, to admit imports from the poor countries, will accelerate the onset of the age of leisure for unskilled and semi-skilled workers. But if all the rich and poor nations can be regarded as a single unit for the cultivation of technology adapted to planetary needs, the 'class distinction' between rich-oriented and poor-oriented technology may eventually disappear. The earliest effect would be on scientific and technological research and on the emphasis within the more versatile technologies such as nuclear energy, satellites and ocean engineering.

For example, more attention would be paid to medium-sized nuclear reactors better adapted, economically, to the needs of a poor country than the giants now favoured in the rich countries. Desalination of seawater using nuclear energy already figures as declared interest of the leading nuclear nations, of potential benefit to arid and semi-arid countries, but the efforts are probably not yet commensurate with the importance of the undertaking. The development of multi-lingual direct broadcasting satellites for educational purposes is at least as exciting a challenge as other space activities. There is tremendous scope for using satellites observing the Earth, not only for meteorology and espionage but also for survey of natural resources in the poor countries.

Particularly desirable would be the more positive diversion of the technologies and advanced skills of the rich countries towards material aspects of the Welfare World. This theme is one which requires much more study if problems are to be identified which will arouse the professional enthusiasm of men accustomed to dealing with rockets and big computers.

The UN advisory committee on the application of science and technology to development made a start in trying to identify particular areas where new knowledge, and therefore special research and development efforts, are needed. Some of them are well worthy of attention from leading engineers. For example, large flows of water, feeding wells, occur deep underground and are difficult to measure and map. In hot, arid areas this underground water is safe from loss by evaporation. Advance in techniques for dynamic surveys of underground water sources will open the way to the construction of underground dams, of kinds never attempted before, to regulate or guide the flow of water.

The UN advisers invited the automobile engineers to develop

vehicles especially suited to use in rural conditions in the poor countries, where roads are poor or non-existent. Hovercraft are among the options here, but one also thinks of the strange rolling, walking and crawling vehicle developed as a possible way of moving about on the Moon's surface. Again, unconventional sources of energy merit more sustained investigation; the harnessing of the solar rays to economic power production in the poor but sunny countries is a major technological challenge, which may find its solutions in novel chemical engineering or in solid-state electronics. These are just a few of the subjects identified, by the UN committee, where acquisition of new knowledge is badly needed.

The habitual and historical technologies of the rich countries are not necessarily appropriate for poor countries developing in a new and more abundant technological setting. There is every reason why the poor countries should attempt to leapfrog over the obsolescent technologies of the rich. If it will take one hundred years for the poor to catch up with the rich, they are aiming at a moving target; if they finish up in 2070 with conditions like those in the USA or Europe in 1970, they may find the rich countries have moved far on to something different and better. In any case, imitation of the present rich countries may be quite inappropriate for countries with different climates, cultures and interests. The poor countries must therefore formulate their own visions of the future and experiment with novel technologies themselves—for example, new forms of surface transport, new ideas in building and town planning, new ways of producing food.

To do that, they will have self-confidently to create their own brands of science. The emphasis in the science of the rich countries need not be imitated indefinitely. It may be that biology rather than machinery is the key, not only to the escape from poverty, but also to the invention of new styles of rich life. In a generous collaboration with the poor countries, in work seeking novel solutions to ancient problems, the rich should learn much of benefit to themselves.

Research and technology are not sufficient for creating a

Welfare World; the political, economic and social adjustments are at least as necessary and in many ways more difficult. But, beyond any reasonable doubt, the re-orientation of research and technology in support of such a programme would release a torrent of humane and ingenious enterprise that would multiply the benefits of political advance. Given the prospect of a century of creative work to a self-evidently desirable end, the enrichment of the whole world, the present habit whereby research and technology attend to the needs of the poor only as an afterthought is hard to tolerate, let alone defend.

Part IV

DEMOCRACY RENEWED

INHERITED political ideas and parties are in disarray, now that the traditional radical-conservative conflict is over-ridden by the radicalism of technology. Meanwhile, computer experts, technological fortune-tellers, anarchic students and various other groups are potential bidders for political power. Yet the reasons why a science-minded power group in charge of a nation would feel obliged to consult public opinion, and why systematic attempts to forecast the future encounter a methodological impasse—these provide a new justification for democracy in the scientific age.

A power vacuum exists, in the failure of existing institutions to look ahead other than in a piecemeal fashion. The prime problem in social control of the uses of science is to reconcile specialist expertise and long-range planning with the generalism of democracy. It must be done in such a way that the wishes of the ordinary citizen are heeded and the experts neither dictate nor bow to administrative government. The only solution is to bring experts and the public face to face in a continuous debate about goals. The eventual vehicle will be the universal university, of which everyone is a lifelong member.

In the meantime a futures debate using electronic and other media will serve as the forerunner of this 'democracy of the second kind'.

As a simultaneous development, political conflict will swing away from the obsolescent radical-conservative axis of dispute. Opinion will regroup as the technological opportunists (Zealots) divide themselves from the scientific conservationists (Mugs) in accordance with a basic difference in attitudes to the uses of science. It is fairly clear which side must win, in the long run.

CHAPTER 14

Dissolute Parties

THE Bonnington Hotel in Bloomsbury, London, was the venue for the most serious attempt ever made in the West, to involve experts in the development of a party science policy. It was there that a senior Labour politician, Richard Crossman, acted as ringmaster for a circus of research workers, engineers, and industrialists to develop Harold Wilson's theme of the technological revolution into a coherent and sophisticated programme for the Labour Party, in preparation for the 1964 election.

The prehistory, from the days when Hugh Gaitskell was leader of the Labour Party, was of a small and dedicated group which, before and after the 1959 election, had developed policy ideas and produced a report on *Science and the Future of Britain*; it envisaged, among other things, the creation of a Scientific and Technical Planning Board in government. Although one effect of the Labour Party activity was to provoke the Conservative government into appointing a minister for science, Hugh Gaitskell was scarcely interested in the quest for a science policy.

Not so Harold Wilson, who became leader after Gaitskell's death. He entrusted Crossman with evolving a science programme and testing it by criticism from experts. In these meetings, the idea of a Ministry of Technology emerged, although there was considerable uncertainty and disagreement about what form it should take. In any case, the actions of the Conservative government of the day, in overhauling the administrative machinery concerned with science and technology, complicated thinking about departmental organization.

In the 1964 election Labour made some play with science and technology and, although it never really became a major election issue, many science-minded people were excited about the possibility of a science-minded government. After the narrow Labour victory, the Ministry of Technology was duly set up, but academic and non-industrial research was left where the Conservatives had put it, with the Department of Education, and at first no adequate provision was made for co-ordinating academic research and technology. Crossman himself was switched right away into housing and, in what was variously regarded as a political dodge or a stroke of genius, Wilson appointed as first minister of technology Frank Cousins, a trade union leader who at that time was not even a member of Parliament. The two chief experts of Crossman's circus were given high posts: Vivian Bowden became the minister responsible for higher education and science and Patrick Blackett became Cousins' chief adviser. Charles Snow, who had also been consulted on Labour Party science policy since Gaitskell's day, went to the Ministry of Technology, as a parliamentary secretary. Neither Bowden nor Snow lasted long in office; Frank Cousins resigned within two years over wages policy.

High hopes were thwarted in a chronic economic crisis and technological renewal seemed in retrospect an electoral gimmick; in practice it would be slow to start and slower to pay off. Even by the 1966 general election, technology had become a muted theme for Labour. The work of the ministries and their advisers continued but there was little evidence that the growth and implementation of science and technology was much faster or more effective than it would have been under the Conservatives. 'Brain drain' figures and the mutterings of senior scientists and industrialists showed that Wilson's 'white-hot technological revolution' was only lukewarm. Meanwhile, Wilson had adapted the technology slogans for his bid to lead Britain into the Common Market.

Miscellaneous actions in the field of science and technology by the Labour government were not negligible, but the sense of anti-climax was unmistakable. Yet, though the people who had gathered at the Bonnington Hotel would admit to disappointment, few would regret their participation. The same cannot be said of the research workers and engineers who aligned themselves behind the Democratic Party for the American presidential election of 1964. Those who lent their support to the Johnson-Humphrey ticket in 1964 saw their loyalties shattered by Lyndon Johnson's Vietnam policy, which really had little directly to do with science policy.

As in the case of the Labour effort in Britain, there was some prehistory. In 1959, in preparation for the following year's US presidential campaign, a group was formed in the Democratic Party concerned with science policy. It was headed by Ernest Pollard, biophysicist. Some Democrats had awoken to the fact that, with the party then out of office, it was not getting expert advice even partially comparable with that given to the Republican government of the day. Also,to identify some leading research workers and engineers publicly with the Democratic cause seemed electorally useful. The Republicans sought to follow suit. Although they were able (according to the Democrats) to produce very few people actually listed in *American Men of Science*, Richard Nixon had a science brochure when the election came promising, most notably, biomedical research in plenty.

In the Democratic camp, the key Senators—John Kennedy, Lyndon Johnson and Hubert Humphrey—had nothing to do with the science group, which was not regarded as having an official position within the party. The former president, Harry Truman, was particularly keen on its activities and, in drafting of the party 'platform' for the 1960 campaign, most of the group's material was used. But the group became identified particularly with Adlai Stevenson, an unsuccessful candidate for the party nomination. When Kennedy had the nomination, he made little direct use of the proposals in his campaign, with the notable exception of the Pollard group's scheme for a Peace Agency (see Chapter 3). American political parties are in any case vestigial bodies which do not as a rule originate anything much apart from candidates, and the presidential candidate conceives and runs his own campaign.

The roll call of members of the founding committee of Scientists and Engineers for Johnson, for the 1964 election, comprised a list of forty-two of the most distinguished research workers, technologists and science policymakers in the country. Alongside the inventor of the electronic television camera, Vladimir Zworykin, and the discoverer of heavy hydrogen, Harold Urey, there were the technical vice-presidents of Ford, IBM, Litton Industries, Lockheed and CBS, the presidents of several universities, and distinguished members of the Washington advisory community such as Harrison Brown, George Kistiakowsky, Roger Revelle, Warren Weaver and Jerome Wiesner. The basis of the movement was not enthusiasm for Johnson so much as the wish to stop Barry Goldwater, the Republican candidate. The group was more interested in domestic than in foreign policy.

Three years later, Elinor Langer (1967) reported her check on current attitudes of the members of that founding committee. She found a strong anti-Johnson group, including some of the most influential members of the original committee. A few dedicated Democrats among the group declared themselves satisfied with Johnson's performance, but there was widespread unhappiness even among those who were not positively anti-Johnson—the main grievance being the Vietnam War.

One of the founder members had openly quit—James Gavin, president of the Arthur D. Little company, who had not only complained publicly about the conduct of the war, but also resigned from the Massachusetts Democratic Advisory Council. Some of the others confessed to Langer their 'timidity about publicly speaking out', confirming the fears of those who had said that the high level of federal support for university research would create political inhibitions among academics. One elder statesman of science was reported as saying that a role in government advisory circles created some restraint; a university administrator was fearful that speaking out about the war might have adverse effects on his institution's relations with Washington.

At the end of 1967, however, an impressive group of Massachusetts academic and industrial scientists launched Scientists and Engineers for McCarthy, in support of Eugene McCarthy's nomination as Democratic presidential candidate for 1968, in opposition to Johnson. That was before Johnson's declaration that he would not seek re-election. The group included many former supporters of Scientists and Engineers for Johnson.

At the height of McCarthy's contest for the nomination with Robert Kennedy, before the latter's assassination, McCarthy had so pre-empted campus support that neither Kennedy nor Hubert Humphrey attempted to form scientists' groups, merely hoping they would inherit the Scientists and Engineers if McCarthy dropped out. Meanwhile, McCarthy went further, and appointed a scientific advisory board, including five Nobel prizewinners and Eisenhower's former science adviser, George Kistiakowsky.

About half of the McCarthy men went over to Humphrey when he won the Democratic nomination. Scientists and Engineers for Humphrey-Muskie was a luminous muster of Nobel laureates and academicians more numerous than the 1964 Johnson team. For the Republicans, Lewis Strauss, former chairman of the Atomic Energy Commission, gathered a much stronger group of scientists and engineers in support of Nixon than he had managed for Goldwater in 1964. It was headed, predictably, by Edward Teller and Willard Libby, among the very few outstanding scientists persistently associated with the political Right. But they were joined by many other academicians—only about a third of the number that Humphrey drew, to be sure, yet a distinguished enough bunch.

Party politics is a portmanteau of disparate people and policies, and the British and American experience suggests that free-thinking academic experts cannot be expected to approve all features of a party's programme or to remain loyal whatever happens. Certainly, experts should be prepared to inform and advise politicians just as they would governments, but the onus for exploiting such advice, and for formulating and propagating policies, rests firmly on the politicians themselves.

In future, the natural framework for science and technology in party political thinking is a general picture of social goals—what will be, and should be, happening during the next decade or two. The uses of technology will typically form an important but not exclusive part of that picture. Such long-term thinking, well done and presented, should certainly interest and impress the electorate, so that rival parties may come to try to outdo one another in their visions of the future. Austria, scientifically one of the least developed nations in Europe, was the scene for an initiative of such a kind, when the Volkspartei (People's Party) drew up a broad twenty-year forecast, *Aktion 20*, for the 1966 election campaign. It provided an initial sketch of developments in education and science policy, law, public health, international relations and the national economy. In Sweden, not long afterwards, the Högerpartiet (Conservative Party) was engaged in a study of the future.

Gustaf Delin, research director at the Högerpartiet headquarters, held that future developments would alter the basic concepts of politics for all parties. An important inner circle of the party, including the relatively young leader, Yngvy Holmberg, showed enthusiasm for this line of thinking. As a consequence, the Conservative Party became reconciled by its own studies to the inevitability of more planning in the nation's affairs; it also identified environmental pollution as a new political issue of the first order. On the other hand the studies did not, at least at first, reveal any special basis for party differences about the future. Indeed, Delin thought that if the Social Democrats undertook similar studies they would naturally arrive at similar conclusions.

PARTY PORTFOLIOS OF TECHNOLOGY

Is that necessarily true? Do political differences disappear when projected into the future? Must parties with different social visions 'naturally' share in a consensus of technological necessity? If so, we really may be on that technological railway (Chapter 5). Research and technology, important though they be as determinants of future life, would then fall outside the domain of political controversy, except at the level of arguing about how to travel faster on the railway. Look around the

Western world and you will not find much that is distinctive in political parties' policies for science and technology, beyond proposals for 'more science' and organizational reform. There is little sense of the opportunity to build a chosen future from available and sought-for elements. In an illuminating comparison, Stevan Dedijer (1964) found that the science programmes of the Republican Party of the USA and the Communist Party of the USSR were virtually identical.

Some of those who are anxious to safeguard research against Lysenkoism and technology against inefficiency swear that these matters are above and beyond party politics; they deplore party differences about ownership and organization. In its first report, the British parliamentary select committee on science and technology recommended policies for the reorganization of the nuclear energy industry. It proposed a centralized, 'socialist' solution, involving the creation of unified national companies for nuclear boilers and nuclear fuels. Naturally enough, Conservative opposition members voted against the scheme in the committee, though they fully concurred with most of the other conclusions about the industry. *Nature* (1967) expressed horror at such a result:

'The committee has done its reputation a great disservice by coming out with a report which is crudely divided on party grounds. . . . Its failure to agree on proposals for the reorganization of the civil nuclear power industry is a great setback.'

The attitude here revealed by the world's most celebrated scientific journal reflects both the scientific man's frequent mistrust of partisan politics and the muddled idea that anything to which the label 'science' can be loosely applied is no fit subject for intrusions of partisan opinion. Somehow, it seems to demand, parliamentary democracy must take an interest in technical matters but not in the context of the party rivalries by which that democracy works. With less orthodoxy but greater plausibility—having insulated the actual conduct of research from possible political deformation—one can argue that putting the results of research to practical use should be a more political matter, in the partisan sense, than it is now. Short of that, it should at least be much more controversial, with forceful and open expression of conflicting views.

Rarely, a political party has become identified with a particular technology. For example, a working-class party may be obliged to be sentimental about coal mining. And long after the era of Franklin Roosevelt and the creation of the Tennessee Valley Authority, the US Democrats were the party of the big dams. It brought them into conflict with conservationists who did not want dams, certainly not in important unspoiled places like the Grand Canyon. Incredible though it may still seem to those who think of the Grand Canyon as one of the natural wonders of the world, water authorities in the Colorado basin proposed, in the mid-1960s, the construction with federal aid of two dams, Hualpai Dam and Marble Canyon Dam, the former to be sited in the national park itself. The Sierra Club and other conservationist bodies opposed the idea and in 1967 Stewart Udall, secretary of the interior, announced that the federal government was no longer supporting the plan. The Democratic congressman Wayne Aspinall of Colorado continued to fight for the Hualpai Dam.

At the Bonnington Hotel in 1964, there was an unsuccessful effort to swing the Labour Party science discussion from generalities to particulars, from institutional arrangements to the items of research and technology that a Labour government might wish to emphasize on political grounds. The attitude of the chairman was that such things were a matter for expert investigation once power was achieved.

In the same year, at the science and parliament meeting in Vienna, I sought to persuade the politicians that they should modernize their ideologies with suitable admixture of specific technologies. If politicians had different views of how the nation and the world should be, they should surely differ in the uses they wished to make of science. Policymaking committees of the political parties, with the blessing of the party leaders, should set out to review some of the obvious trends in science and technology, through conservative, liberal or socialist eyes, and consider what tendencies they should favour and what oppose.

Where that was not appropriate, I said, they should ask whether their existing policies need modifying in the light of particular scientific developments. A successful start along these lines would generate a self-sustaining chain reaction of interest,

---

VI AUTHOR'S 1964 TABLE

Conceivable differences of technical emphasis under different policies for an advanced country.

FAVOURING COMPETITIVE PRIVATE ENTERPRISE	FAVOURING NATIONAL OF MONOPOLY PLANNING
Versatile programmed production methods	Large automated transfer production lines
Small technical projects	Large technical projects
Small computers	Large computers
Small power units	Large power units
Better telecommunications and transport	More elaborate weapons systems
FAVOURING RESPECT FOR THE INDIVIDUAL	**FAVOURING SOCIAL WELFARE**
Greater variety of products	Preventive medicine
Psychological research	Social sciences
Ergonomics	Building and urban research
Consumer research	Rural projects
FAVOURING OWN NATION	**FAVOURING WORLD DEVELOPMENT**
Newer technologies	Older technologies
Synthetic products	Natural products
More profitable food production	Greater food production
Weapons	Reproductive physiology
National space programme	Tropical medicine
	Earth sciences

in which parliamentarians, rank-and-file party workers, and sympathetic scientists and technologists would find themselves caught up.

The table prepared for that meeting is reprinted here. Although I described it at the time as a guess, and some modifications might be appropriate five years later, I think it remains roughly valid. There is certainly a sense in which, for example, genuine 'private enterprise' conservatives should vote for fuel cells, as a means of generating electricity on one's own premises, in preference to big national electricity grids based on nuclear fission. Similarly, socialists keen on overseas aid should want to see more research on food production and birth control.

But I am less optimistic now about parties developing coherent science policies on a basis of portfolios of preferred technologies. The lesser reasons are practical. The interconnections of technologies make political logic difficult: for example, those 'private enterprise' fuel cells might have to be supplied with natural gas by a state-owned industry; again, efficient communications, necessary for decentralized, small-scale operations, will in future depend on big computers. Another practical objection, which I took more fully into account at Vienna, is that technical circumstances change quickly and politicians would have to be prepared to review their opinions continuously and avoid becoming dogmatic—unless the technologies, because espoused, became ends in themselves.

A more important reason for doubting the possibility of apportioning technologies to parties is that, in practice, the attitudes that serve as headings in Table VI, while still very relevant for human affairs, no longer correspond closely with the actual policies of political parties. When political ideas are in disarray, how can one attempt to pin technologies to them?

POLITICAL DISARRAY

The advanced Western countries in the 1960s were so much in the grip of technological nationalism and its compulsions that there was consensus among the political parties. A planned economy of mixed public and private enterprise, dedicated to continual economic growth and effective application of new technologies, seemed to be to everyone's taste. There had to be big investment in education and research. Other commitments permitting, social welfare programmes would avoid the grosser effects of domestic poverty, and aid would go to the poor countries if there was anything to spare.

Add to these goals the assumption that 'scientific' government, based on better and better information, would lead to what Asa Briggs (1965) called 'an unconditional surrender to the facts', and a belief that expert advisers would provide definitive recommendations about the uses of technology, and the recipe seemed complete for the virtual abolition of serious political conflict. For many, such a triumph of rational consensus was greatly to be wished. But were the goals of the consensus necessarily right?

Workers who voted for social democratic parties found their trade unions being bent into the planning framework, and themselves being thrown out of work by technological advance. Old-fashioned Conservatives who voted for competitive enterprise and lower taxes found their governments pouring huge sums of taxpayers' money into selected companies, for unpleasing novelties like government computers and supersonic airliners. By the consensus, objectors to such trends were obsolete. But the young were hardly obsolete, and the prospect opening for them was a materially richer world, certainly, but also one lacking much sense of purpose, and where it was hard to see a career pattern, let alone a vocation. They revolted, in country after country.

Others who were well-informed about technological trends were appalled by the prospects of unending contests in space and the deep ocean. They foresaw wild animals and plants banished to the museums, to end their intrusion in a man-made world. They feared computers that would not only calculate how it was most profitable to live but would be watching every citizen's action, transaction and misdemeanour with a brotherly eye, lest idiosyncrasy and nonconformity frustrate the quest for mechanical wealth and national power.

In the face of such dismay, the consensus cannot survive, but it is unlikely that the old political ideologies will prove to be satisfactory vehicles for the controversy. On the Right, traditional conservative ideas cannot be sustained now that so much of research and technology for industry requires state support; now that, in a world committed to change, the stability that has to be conserved is more like the stability of a jet aircraft than the stability of a castle. Traditional parties of the Left are particularly vulnerable both to the consensus and to the direct effects of technological change. The first is plainly destructive of the parties' ideology. The other erodes their sources of power among industrial workers, by absorbing skilled workers into near-managerial status in science-based or automated plant and threatening others with large-scale redundancy.

In time—not long in the rich countries— automation and the production of great material wealth more or less at will are going to undermine the basic economic values of goods, services and labour. Work and wages as understood today may become simply obsolete concepts. How will society be organized in a highly automated age? In the USA, a national commission on technology, automation and economic progress recommended that Congress should examine wholly new approaches to the problem of income maintenance and give serious study to a minimum income allowance or a negative income-tax programme. The commission's purpose, in this recommendation, was to make it easier for people to adjust to 'a fast-changing technological and economic world' without major interruptions of employment.

Herbert Hollomon (1967a), at the time the leading scientist in the Department of Commerce, predicted that in the 1970s the entire American political scene would be turned topsy-turvy by the fact that more non-workers than workers would be voting at elections. Within twenty years, the average American can cut the amount of time he spends working for his living by half, if that is what society chooses; otherwise, working at the same level, the United States could embark on a colossal foreign aid programme, giving away $21,000 million by 1970 'without even noticing it' and increasing that amount by $7000 million every subsequent year.

'Already, America is producing the first generation of children in history who do not have to worry about earning a living,' Hollomon said. 'These kids are the sons and daughters of kings. They are surrounded by more luxuries than the monarchs of England were in the eighteenth century. Work no longer dominates their horizon and they sense even now that much of their education is irrelevant. That is the psychological gunpowder behind today's student rebellion.'

A less cheerful view of the same phenomenon was taken by the US secretary of labour, Willard Wirtz (1964). If a human being was to compete with modern machines, he had to possess at least a high school diploma. Yet the trends indicated that as many as thirty per cent of all students would be high school 'drop-outs' in the 1960s. Wirtz said: 'The confluence of surging population and driving technology is splitting the American labour force into tens of millions of "haves" and millions of "have-nots".... This division of people threatens to create a human slag heap.'

THE REGROUPING

Contrary views of automation, sanguine and fearful, can broaden out to general attitudes to the whole field of technology. I believe that they represent, far more closely than any traditional party polarity, the necessary political dialogue of our time. These attitudes, among politicians and public alike, cut across the conventional party lines. In many cases, conservatives have become radicals and radicals conservatives. Those who scoffed at a scientific view of society and the idea of a brave new world inspired by research and reason have now seized on science-based technology as a means for military and economic aggrandizement and as the great mover in human affairs. Middle-road, meritocratic people dedicated to economic growth as the primary goal are delighted with technology and can hardly have enough of it. On the other hand, people committed to social reform are now often the first to be dismayed by technology, and may find themselves joined by nostalgic conservatives, romantic liberals, and workers fearful of technological unemployment. Open resistance to change is still relatively slight; it is a brave public figure who is opposed to technical progress in general. Nevertheless, among many ordinary people the misgivings are strong indeed—as they are among some of the research workers themselves.

In the 1930s and 1940s, a legitimate generalization was that research scientists in Europe and North America were radical and engineers were conservative. The typical research worker was on the political Left; he nurtured optimistic dreams of a New Jerusalem built by the application of scientific ideas, and he resented the engineer's affiliation with private industry. The engineer, for his part, was on the political Right; he identified himself with the company for which he worked and was unconvinced about the revolutionary possibilities of new discoveries. Now the situation is at least partially reversed. The engineer has the bit between his teeth. What the research worker was saying about the practical implications of his work has turned out to be correct and the engineer and his company are hell-bent on change. The research worker may now be unconvinced that change is necessarily for the better.

Articulate scientists are polarized into the optimistic 'radicals', or 'fixers', and the pessimistic 'conservatives', or Cassandras. Typical of the former is Walter Orr Roberts (1966), director of the National Center for Atmospheric Research at Boulder, Colorado:

'It is here, this age of science! And with it comes its promise of what the life of man can be, with food for all, with education, with human freedom, with a stable population. With this age of technology comes the realistic expectation that even the added billions can live in harmonious equilibrium with a natural environment of quality. . . . Rather than simply fight for the preservation of the old things that are good, we must plan creatively also to shape the new. We must commit ourselves to dare to build the world we want, knowing that it is possible if we but demand it—and if we use intelligently all the potent forces of science, the arts and the humanities that are at our disposal.'

Contrast these with the words of the professor of botany at the Washington University, St Louis, Barry Commoner (1966): 'Despite the dazzling successes of modern technology and the unprecedented power of modern military systems, they suffer from a common and catastrophic fault. They provide us with a bountiful supply of food, with great industrial plants, with high-speed transportation, and with military weapons of unprecedented power—but they threaten our very survival. . . . Technology has not only built the magnificent material base of modern society, but also confronts us with threats to survival which cannot be corrected unless we solve very great economic, social and political problems.'

Is it along these lines that political parties will regroup? Or are these basic differences of outlook about our technological civilization to go unrepresented in the party political arena? The views are what I shall for the time being call Z and M:

Z. *technological opportunism*—a dedication to an imaginative leap into the future on the basis of material progress.

M. *scientific conservationism*—a commitment to the preservation of human and environmental well-being in spite of material progress.

Although one is broadly 'radical' and the other 'conservative', in the general meaning of those words, the interests and ideals represented cut right across those of the traditional parties that use the labels.

Party Z, in the technological context of the early 1970s, might emphasize the development of a total information system based on a computer-communications network and huge electronic data-banks; major reconstruction of surface transport facilities; a substantial investment in the development of factory-made foodstuffs; a crash programme in spare-parts surgery; big new enterprises, civilian and military, for the conquest of space and the deep oceans, preferably without constraint by international obligations; and the ordering of 'far-out' prototypes such as a Mach 5 hypersonic airliner and a giant 'intelligent' computer. In scientific research, it might favour massive, co-ordinated programmes aimed at synthesis of life-like chemical systems and at weather control.

Party M, the scientific conservationists, would prefer those technologies which bore most closely on welfare, like agriculture, building and crime prevention. Public educational systems using advanced communications and computing technology would be promised. Preventive medicine would be preferred to heroic surgery but on the basis of voluntary rather than compulsory measures; scientific interests would take precedence over exploitation in atmospheric, space and ocean research; diplomatically, there would be strenuous efforts to internationalize the ocean bed. Party M might also promote centres for 'human studies', bringing together social and natural scientists for a major development of social sciences along fresh lines. Its environmental policies would hinge on pollution control and on protection and restoration of the countryside and city centres.

To be practical, even if modern communications accelerate the political process, a brand-new party is unlikely to be relevant to the problems of adjustment to technological change in this century, let alone in the 1970s. Existing political parties might more quickly change their complexions so that they tended to pair off into positions corresponding approximately to Z and M. Even to expect that is optimistic. But controversy in other arenas, outside the parliamentary party system, may more readily polarize along Z-M lines.

CHAPTER 15

Fixers and Forecasters

BEFORE Chaim Weizmann, chemist and microbiologist, became the first president of the new state of Israel in 1949, he dreamed of developing a really good process for converting cheaply grown starch into alcohol by fermentation, and then into butyl alcohol and acetone. 'These three materials, in large quantities and at low price,' he said, 'could form the basis of two or three great industries, among them high-octane fuels, and would make the British Empire independent of oil wells.' (Weizmann, 1949)

Like other Zionists, Weizmann had found anguish enough in the relationship of oil with world politics; hence his wish to break oil's monopolistic position. In the Middle East, where Jews were to establish their home, the major powers were preoccupied with oil supplies from Arab countries. Feudal and militaristic governments, sustained by Western powers in that area, assured supplies of the liquid fuel. Within the geologically lucky countries—which did not include the Promised Land—vast royalties from oil wells went to the rulers.

Were Weizmann still alive, he would surely feel confounded. Oil remains a fuel easily transported, and made infinitely variable by distillation, cracking and blending. Road transport, sea transport and aviation still depend upon various fractions of petroleum—and so, now, do the Moon rockets. Unlike the coal industry, the oil industry is unperturbed by the rise of nuclear power. Its only long-term worry is whether it can find enough oil to meet the other rising demands but, with the prospect of the recovery from oil shales and tar sands, it has plenty to be getting on with. When oil and natural gas were already the source of just over half the world's energy, Monroe Spaght (1966), a managing director of the Royal Dutch/Shell group, forecast that the fraction would grow to 59 per cent in 1985. By the end of the century the world would be consuming energy four times as fast, so the oil industry would have to find more oil and gas fields than it had discovered since its origin in the nineteenth century.

Yet the petroleum industry's most rapidly growing market was in activities other than energy supply—in particular in petrochemicals, a major technological sector of all advanced countries. According to Spaght's forecast, by 1985 oil and natural gas would become the source of virtually all organic chemicals and, on a world-wide basis, a twenty-fold increase in organic chemicals was to be expected before the end of the century.

Weizmann's proposal for using crops to make gasoline was stood on its head. In the 1960s the oil companies began to use petroleum to grow food. They found yeast suitable for feeding to farm animals, perhaps eventually to humans, which would grow using the chemical energy of the waxy fractions of oil, instead of the sunlight needed by conventional crops.

Weizmann could anticipate the course of research and technology no better than anyone else. His experiments, sponsored by the British government, were of no avail. Had they been—if we can ignore for a moment the ecological problems of gathering the necessary starch—he might indeed have done down his Arab enemies. Whether you take his chemical work for Britain or his dreams for Israel, Weizmann was an outstanding example of what would now be called a 'fixer'.

The chief prophet of the technological fix is Alvin Weinberg, director of the Oak Ridge nuclear laboratory in the United States. His doctrine, of the use of knowledge and skills in a very direct manner to solve visible social problems, expresses the sense of power felt by many research workers and engineers. Weinberg, in 1967, was himself the man behind the Eisenhower plan for agro-industrial complexes—a technological fix for the Israel-Arab confrontation.

In the near future, nuclear reactors and advanced desalting techniques should be able to produce fresh water at about fifteen US cents per thousand gallons, along with electric power at 0.3 to 0.4 cents per kilowatt-hour. Given efficient, high-yield cultivation, two-and-a-half cents' worth of water should be enough to produce wheat or rice to feed a man for a day. For the first time it may be realistic to think of using desalted water for agriculture, as opposed to industrial and domestic applications. If the electric power is used in part for the electrolysis of water to produce hydrogen, a route to ammonia for fertilizer is opened up—and to other chemical activities, too (Weinberg, 1967b).

This was the technical background to an American proposal for three 'agro-industrial complexes' in the Middle East, to provide an economic and agricultural base, primarily for the rehabilitation of the Arab refugees. Other coastal desert regions in the world could benefit from the same kind of programmes. The cost of each complex would be of the order of $1000 million and it would be capable of feeding about two million people. The plan visualized the creation of a stock company to which individuals and governments would subscribe. At the same time the expenditure would represent a big investment for industry in the poor countries and, if conducted on a strictly commercial basis, should give a good return on the capital, albeit after a number of years. It was also possible to envisage the creation of such agro-industrial complexes on a gift basis, as aid to world development. The addition to the world's population in the mid-1970s will be about seventy million a year, so $5000 million a year invested in, or given for, the creation of such complexes, could supply one-seventh of the additional food supplies. This is not the place for a technical or economic critique of the concept, though there are obvious questions such as the effect of massive chemical production on world prices of the chemicals involved. It is the concept of the concept —the fix—that is of interest here.

Weinberg (1966) placed a very broad interpretation on the technological fix. The development of the technologies of energy and mass production in the capitalist countries was, for him, the fix that enabled capitalism to curtail poverty and inequality without the social revolution that Marx predicted. He also allowed that Hitler's machinery for the extermination of the Jews was itself a monstrous kind of technological fix. And, although he admits that not all would agree, Weinberg sees the H-bomb as 'the nearest thing to a technological fix for large-scale war', meaning that the existence of the H-bomb has helped to prevent war. More generally, Weinberg's fixes are partial solutions which can allow time to get at the root causes of social problems. The advantage is that technological inventions are more readily accepted than a social invention. 'It was easier,' Weinberg observes, 'to start on the atomic bomb than to modify the income tax laws. The beauty of a technological resolution of a social problem is that technology doesn't have deep semantic overtones and connotations.'

Examples of the fixes that Weinberg offers vary in plausibility: the intra-uterine contraceptive device or 'Loop', for the global population problem; desalination of water by nuclear power for the shortage of fresh water; the 'safe' automobile for the road-safety problem; air-conditioning for the summer race riots in American cities.

Weinberg even has a fix for the social problem of engineering technological fixes. He has called for the creation of 'national socio-technical institutions' for the development of coherent doctrines for technologies and public policies. They would bring together scientists, technologists, systems analysts and social scientists within one institution, to work out technological means of achieving social ends while taking into account the social consequences of those technical means themselves. To those who object that such institutions would be far too powerful as originators of social doctrines–as the RAND Corporation is feared to be in the military field—Weinberg replies: 'Establish in each case not one but rather two competing institutes that will keep each other honest.'

The technological fix is a useful idea, but liable to be socially naïve and politically risky. The very fact that a leader of American research can propose such an approach is an indication of how much progress we have still to make in the political control of technology. Desirable though a fix may be in a particular case, it represents an arrogation of power by the technologists who originate it.

Weinberg's fixes deal with pre-existing social difficulties and goals. What I shall call technological 'lures' represent brand-new goals like travelling to the Moon, which could not be contemplated until the necessary research and invention had ripened. Driving fast cars for fun, or skin diving with gas bottles, or collecting pop records, represent essentially novel additions to human opportunities—lures at the personal level. But rich countries since 1945 have specified several major new activities so clearly, and have supported them on such a scale, that we can count them as lures at the national level.

The technological and political content varies, but prominent among the lures are: technical-cum-economic aid to the poor countries; automation and computerization; build-up of higher education and construction of huge research facilities; exploration of space; exploration of the undersea. I should add nuclear deterrent systems as a lure, although Weinberg claims them as a fix. Together, these activities already represent a substantial part of the life of the rich countries. None of them would have made much sense to politicians in the 1930s, but today they are taken almost for granted; even if there may be haggling over budgets sometimes, it is eccentric to wish to abolish them.

We can anticipate further political innovations of technological origin. Paralleling the technological fix, the pure technological lure is a coherent programme—like manned spaceflight—that can be sold as a package to the politicians. Computerization of education and biochemical monitoring of the health of the whole population would be typical lures for the 1970s.

Just as Weinberg noted the ease with which technological fixes could be introduced, compared with social inventions, so we can see that most of the lures mentioned were accepted with remarkably little opposition. The fewest difficulties arise when the purposes of fixes or lures lie well outside everyday human affairs and the traditional preoccupations of the politicians

—except that anything involving international competition also tends to have an easy ride with the politicians. If we intend to look for new fixes or lures that bear closely on the lives of ordinary people, we should expect (indeed wish for) more political controversy. Total reform of surface transport systems, for example, would involve many industrial and local vested interests, quite apart from questions about the political connotations of the chosen system. Moves to develop new facilities for the 'age of leisure' might evoke fierce debate about the purposes of human existence and what kinds of things we should be doing with our time—puritan versus hedonist. An intensive development of non-agricultural sources of food as a fix for world hunger would not go without objections, while the abolition of agriculture which might follow from it could be an unusually warm political issue; in many countries, besides the direct opposition of the farmers, a profound alteration in political and economic geography would ensue. Similarly, vigorous and really effective measures for conservation and restoration of the biological environment would entail highly controversial interference with long-accepted human rights to devastate landscapes.

Indeed, if anyone still bothers to wonder why so much effort goes into reconnaissance of the Moon and planets when there is so much to be done on Earth, one reason is that the biologists have not yet offered competitive fixes or lures for terrestrial problems (perhaps because of their preoccupation with 'advanced' laboratory research). Another reason is that it is politically much easier to be irrelevant.

Unless proposals for future goals are indeed highly controversial, they are likely to be skirting the issues of vested interest and underprivilege, of regimentation and individuality, of opportunism and destruction, which plague us. If technological fixes and lures aimed to court controversy rather than dodge it (as Weinberg would have it) then they might indeed serve great purposes. But they must be seen as political weapons, and therefore as potentially dangerous to someone else. Weinberg was wise enough to recognize the Nazi gas-chambers as a black fix.

In the hands of the political assassin, plastic explosive is a very literal kind of fix. For an industrial proprietor, particular modes of automation might provide a fix against a militant trade union. For a nationalist rebel in a poor country, herbicides may yet provide a fix for the feudal planter, while the latter's friends in the rich countries prepare to fix the rebel with similar agents. In short, to talk of technological fixes and lures begs all the obvious political and ethical questions of motive and interest.

A citizen of the rich West may feel confident about older and blander social goals. During Eisenhower's administration a commission on national goals itemized nine concerns for the American people: the status of the individual, social equality, evidence of the democratic process, more education, economic growth, technological change (yes, a goal in itself!), attention to agriculture, better living conditions, and health-and-welfare. A worthy list—but how conventional, how dull, how vague, how inadequate for men who have stolen the fire of the Sun and do not know what to do with it!

The Soviet regime, too, knows its broad social goals well enough. There are three targets: creation of the material and technical basis of communism; development on this basis of communist social relations; and the making of 'a new man'. The material and technical basis is to come from an immense development of productive capacity in an attempt to ensure that the Soviet peoples have the highest productivity in the world. This target has some explicit technological connotations—in electrification, mechanization and automation, industrial chemistry, general development of new products and new sources of energy and materials. The development of communist social relations is a matter of raising living standards in the rural areas to match those of the city, and to cause the differences between manual and mental workers to vanish with high standards of education for everyone; then the socialist state will evolve into one of 'communist public self-government' in which participation in social life will become normal behaviour and most public duties will be performed by citizens in turn, with participation on equal footing in the discussion and solution of all social problems. 'The new man' that the communists have in mind is an all round citizen combining spiritual versatility, high integrity and physical fitness.

The trouble is, of course, that both these sets of goals are so general, and conceal so many practical issues, that the governments of the countries concerned have to make continuous appeal to other norms for the conduct of affairs, and noble ends are frequently pursued by squalid means. Public affairs have always been like that, but the vague systems of ethics involved are put under even greater strain than in the past by scientific innovations, from nuclear bombs to expensive medical procedures.

THE FORTUNE-TELLERS

Just before Easter 1968, two dozen Europeans sat around a big table in the palace of the Accademia dei Lincei beside the Tiber, to start planning the future of the world. Aurelio Peccei, the Italian industrialist, had summoned leading managers, scientists, social scientists, philosophers and planners of eight nations to start the Western European contribution to Project 1968. This was a scheme for involving experts from the 'scientifically and organizationally more advanced nations' of East and West in an effort to anticipate, by research, world trends, dangers and opportunities, and to plan accordingly and in concert. In explaining his idea to the Russians, Peccei (1967) had called for an act of political will: 'that of deciding that the moment has come to launch jointly and with the co-operation of other nations a project of exploration of the future in order, if possible, to define certain coherent, feasible *worldwide objectives* which mankind should accept for the next ten or fifteen years'.

Ten years earlier, such a meeting would probably have been impossible. Most people would have laughed at the idea. But during the 1960s serious, non-fictional study of the future had become widespread and 'respectable'. This remarkable intellectual phenomenon had plenty of diverse roots: in the writings of Jules Verne, H. G. Wells and their successors; in the needs of military systems analysts and industrial planners for informed guesses about the battlefields or markets of the near future; in the concern of some scientists about the impact of new techniques.

Systematic technological forecasting began quietly in various American laboratories during the 1940s and 1950s. On the social science side, Bertrand de Jouvenel, the political scientist, started the Futuribles project in Paris in 1960. Four years later, a hundred eminent scientists and non-scientists of sixteen countries laid the main expectations for the next twenty years before the public in the series on 'The World in 1984', in *New Scientist* (Calder, 1965). Following de Jouvenel, the French government set up its Commission de 1985. The American Academy of Arts and Sciences established a Commission on the Year 2000; groups were established in Czechoslovakia, Poland and the USSR. Privately and publicly, 'technological forecasting' and the broader 'futures research' were snowballing. In a report to the OECD, Erich Jantsch (1967) identified more than a hundred continuous or periodic activities in technological forecasting in thirteen countries that he visited. Until then, few practitioners had any idea of how widespread their activity had become.

Among the major private companies in the USA and Europe that were taking forecasting seriously at that time were ASEA (Sweden), the Bell System, Boeing, Elliott Automation, Esso (UK), General Electric, ICI, IBM, Lockheed, Minnesota Mining and Manufacturing Company, North American Aviation, RCA, Shell, Unilever, Union Carbide, Vickers (UK), Westinghouse and Xerox. General Motors and Du Pont had actually reduced their effort in long-range technological forecasting after an early start but Jantsch estimated that American industry was spending $60–65 million a year on technological forecasting. He also distinguished a hundred different techniques of forecasting, many of them recently invented. At the time of his report, these techniques had at best only a mild effect on the quality of forecasts, most of which continued to be made in a pragmatic, intuitive way.

Some of the substance of current futures research is employed throughout this book. Here, let us look at the future of futures research itself. It is, inevitably, an art rather than a science, even though computers and other fancy techniques are already being applied to it, especially in the United States; these may give a spurious authority to what are essentially hunches. Forecasting, or futures research, is scarcely a profession. Although a growing number of people are justifiably engaging in futures research full-time, they are dependent to an unprofessional degree on the opinions of other experts. Moreover, every well-informed human being potentially has something to contribute to forecasting and speculation about the future.

This art of forecasting human affairs, whether of technology or politics, is at present crude and hazardous. It consists of making non-trivial but plausible guesses about future circumstances. Forecasts have always been implicit or explicit in almost all human activities, even if the implicit assumption is an unconscious belief that the future will be like the present. Explicit forecasting is becoming more widespread because the rapid rate of technological and social change leaves less time for adjustment to change when it occurs: also because the change itself provides plenty of substance for non-trivial but plausible guesses.

THE USE OF FORECASTS

The most obvious purpose for forecasting is as an aid to planning. Cities, transport facilities, hospitals, schools and similar long-lived structures built today will survive into a period when needs and practices may be quite different. Closely related is the need to anticipate side-effects of change, in order to avoid social dislocation (from automation, for example) or environmental degradation (from pesticides, noise, radioactivity and so on), or to anticipate legal, moral and political issues arising from change.

Again, there is the wish to maintain a competitive position, whether between nations in respect of armaments, between nations in respect of exports, or between different sectors or companies of an industry.

With the greatest early efforts in forecasting located in the

US military-industrial complex, it would also be naïve to suppose that forecasting will not be applied to the goals of military and technological nationalism. A nation strives for early development of new weapons systems to secure military supremacy. And if in possession of accurate but secret forecasts of, say, shifts in the pattern of trade, it might negotiate long-term trading agreements immensely favourable to itself. Such reasons alone provide enough motive for putting futures research on a public and international basis, as Mankind 2000 sought to do, at an 'international futures research inaugural congress' in Oslo in 1967.

Formulation of research and development policy at national or laboratory level will be another use of forecasting—either directly through the identification of isolated technological opportunities or indirectly in the light of general policies suggested by forecasting. Forecasts may also be very beneficial to the young individual, if used to amend the present career guidance which seems to assume the professions and prospects will remain unchanged into the twenty-first century; more realistic assessments will take into account changes introduced by automation, computerization, increased leisure and other factors. But potentially the most important use of forecasts will be for conscious shaping of the future, by preferential reinforcement or discouragement of particular trends and innovations.

It is more realistic to expect major technological and social changes than to assume that things will stay as they are. Yet specific forecasts about the character and consequences of changes may be simply wrong. For example, transport experts predict unabated growth in travel, while communications experts say that high-capacity communications channels and devices such as video-telephones and facsimile transmissions will diminish the need to travel, except for pleasure. The two forecasts cannot both be correct in an unqualified way. This example suggests, incidentally, that errors in forecasting will be reduced by comparisons between different fields of activity.

Reduced, perhaps, but never eliminated. Oddly enough, the fact that wholly new discoveries and inventions are almost by definition unpredictable is not the most difficult problem, at least for forecasts over periods of ten to twenty years. There is usually a fairly long time lag between a discovery and its widespread application. Rather, it is want of sufficient imagination and the impossibility of reliably predicting human actions and choices that make the probable errors in forecasting substantial.

Short-term forecasts tend to be dull and obvious, long-term forecasts provocative and unreliable. The period chosen plainly depends on the purpose for which the forecast is intended, bearing in mind that timidity may be as hazardous as excessive boldness. If the aim is to use forecasting as an aid to the orientation of national policies, twenty years is probably the most useful time-scale. There is interplay between forecasts over different periods. A forecast for twenty years hence should take account of what people may then be choosing with an eye to the twenty-first century.

Harvey Brooks (1967) and his colleagues warned that technological forecasting could do more harm than good for research planning if its results were treated as more than rough first approximations. 'Part of the purpose of research is to keep many future options open to society, and the purpose of technological forecasting is primarily to identify and expand the options rather than to foreclose them.'

POLITICS IS INESCAPABLE

Conscientious study of the future will not necessarily lead men and governments to make the right choices and decisions. Futures research, like any other branch of learning, may be misapplied to nationalistic or selfish ends. Futures researchers should not expect that important statements about the future will be enough to make world leaders see what the goals should be and what dangers need to be averted. That would assume that the forecasts are not only sound but also non-controversial, in the sense that it is possible neither to disbelieve them nor to place alternative interpretations on their significance; it further assumes that political leaders should act rationally on the forecasts.

In the case of population growth, people make strictly numerical forecasts about trends on the basis of the long-standing science of demography; their conclusions are reinforced by the visible multitudes of babies. Even so the world's leaders have been very slow to move, either towards control of population growth or towards meeting the economic problems generated by a doubling of the world's population in the decades immediately ahead. To secure action about population control does not require more forecasting on the part of the demographers—except, of course, to keep their estimates up to date. It depends upon persuasion, upon propaganda about the problem, upon securing a foothold in the centres of power for the idea that action is desperately needed. In short, it depends upon politics, with all allowance for irrational, chance and corrupt processes of human government.

Human choice remains a crucial and unmanageable factor in any practical forecasting activity, and the principal large-scale medium of choice is politics. Drawing a contrast between two different kinds of forecasts helps to illustrate the inevitable political content of forecasting.

Distinguish, then, the 'exploratory' and 'normative' modes of forecasting. The first explores the possibilities arising from current trends and innovations, by extrapolation. For a simple example, graphs of recent progress may suggest that passenger aircraft will get larger or faster (technological forecast) and passengers more numerous (social forecast). But present possibilities for technical and social change exceed the capacity of human institutions to implement them all. From combinations of selected possibilities, a multiplicity of possible futures can be constructed. The farther into the future one looks, the wider the spectrum of choice becomes. For this very reason, unless the exploratory forecaster is blind to the difficulties, he is quickly frustrated by the fact that he cannot tell in which direction the thrust of technology will go, as a result of human choice. So he cannot hope to make a very reliable forecast, unless he can guess correctly what human beings will choose.

Normative forecasting, on the other hand, is concerned to

identify a desirable future, or fragment of the future, and then to track back through the steps needed to accomplish it until the first steps can be reconciled with current possibilities. Ideal communism is an example of a broad normative forecast; the goal of landing men on Mars is a relatively narrow one; dreaming up the ideal domestic cooker for the late 1970s is narrower still. Sensible normative forecasting can only occur within the cone of possibilities suggested by exploratory forecasting but, once the goal or norm has been fixed, an axis of change becomes more or less determined.

Now, a forecaster is entitled to say what he thinks *should* be, as well as what he thinks *will* be. But no individual futures researcher or self-appointed group can seriously fix goals for the whole of society. The forecaster can set down norms and draw conclusions, but it remains an exercise, which cannot lead to reliable forecasts.

Here is a methodological impasse—the booby-trap for the fortune-tellers. Normative forecasts are necessary, but they are political statements. The future is politically charged and politics is, or ought to be, futuristically charged, so that forecasts involve political assumptions and are prone to falsification by political decisions.

It is surely no cause for grief that the future of mankind cannot logically be an exclusive intellectual preserve of self-nominated experts, nor be considered apart from politically controversial issues. Practical, probably institutional, ways will have to be sought, for closing the loop between forecasters on the one hand and the politicians and public on the other.

One proposal for what may be needed came in a philosophical treatise on forecasting by the French pioneer, de Jouvenel (1964). He called for the creation of *un forum prévisionnel*, or surmising forum. It would be a true institution, conceived for the purpose of integrating into more general forecasts the specialist forecasts by experts in very different fields. As de Jouvenel described it, this surmising forum would not only help the decision-makers by making forecasts but would also make public the forecasts used by the decision-makers.

He saw the responsibility for this work falling primarily on the social scientists. Correctly, he took the view that technological change is only one of several factors of change, but to make his point he tended to understate the social impact of technology. He suggested that the social scientist forecasters could learn what they need to know about technological information from semi-popular scientific journals. The creation of a surmising forum is an excellent objective and social scientists may indeed play a leading part in it but, if natural scientists and engineers are not well represented, it may be judged to resemble a car without an engine.

To the political future of futures research we shall return later (Chapters 16 and 17). For the moment, Gianni Giannotti (1968) of Turin, in a report for Project 1968, has words that should be heeded: 'Without a well-grounded concept of our humanity, without a legitimate hypothesis of the trends of our culture and others', every plan is a bet not only against the unknown itself but also against the unknown in us.'

A USE FOR VISION

'Yesterday the party did poorly: today young Pat is hungry, a little shaken, and sorry about his friend James, who met the notorious Samson—not on their list—and was quite badly mauled before the rest of them scared the shark away. A rescue airship came stealthily to take James back to Dublinberg. Pat had asked the pilot whether Matilda was in this sector at all, but the pilot would only say there was laughter in the surveillance room about the party's fruitless searching. They would know on Dublinberg exactly where Matilda was. Even if she didn't pose in front of one of the many camera-buoys they would probably have heard her call; failing that, there would be reports from the management and research parties, who had all the submersibles and technical aids.

'This month Pat is living "rough and wet" as an assistant hunter, swimming by day and sleeping by night with the others of the party on the raft. Under water, Pat breathes through an artificial gill-membrane strapped on his back, so he needs no air bottles or other means of survival, except an insulating skin that keeps out the cold. Pat is not allowed to take any fish himself, let alone a shark; but he has plenty of putative kills in his camera gun. Fergus and John, the two licensed hunters, have to supply all the food for the party, by crossbow-harpoon and hand net, and even they can only catch the species and maximum quantities designated by the managers. In the case of the larger fish, they can take only particular individuals.

'The licensed hunters would not have much chance without the help of kids like Pat, who could see and smell and hear far more. The older men, not born to the life, did not even recognize the sharks but would ask the kids, "Is that Matilda?" when they glimpsed a young bull shark—as if they had not seen the identification films half a dozen times before they left Dublinberg.

'Fergus, like Pat's father, had helped to build the floating city, at the big freezer on the Shannon; during the Abolitions they had been among the last, unwilling to give up work. Even now, except on the long killing trips, they found it hard to live by the environment game—scientific gamekeeping by land or sea, factory-made food to eat most of the time, and the leisurely, studious life under the dome of the city, based on automatic production of everything except designs and ideas. But Pat, born on the ice island, was able to swim and ride a dolphin almost before he could walk, and knew the beasts of the sea as well as he knew the family cat; what to his father was "this marine biology stuff" was to Pat the only way of life he knew.'

That is a fragment of a personal vision, based on a more formal presentation of a possible future (Calder, 1967b) which sought to deal with a range of foreseeable problems and opportunities in a fairly comprehensive proposal for life in the twenty-first century. It faced the questions of how to feed the much larger human population without destroying residual wild life; of how society might be organized when automation had eliminated industrial work and made obsolete the conventional systems of economic reward; and how to take account of man's biological nature as a hunter. The result was a compound scheme, involving: (1) factory production of food and abolition of agriculture; (2) redistribution of the population evenly over the globe by land and sea, in compact towns where the most advanced technologies would be available; (3) restoration of the remainder of the environment to something like natural conditions; and (4) invention of an elaborate gamekeeping system that would provide the principal 'occupations' for humans, as scientists and hunters.

What is the use of visions like that? No individual's idea is ever likely to command faithful and universal acceptance. They cannot conveniently design (or prescribe) all aspects of life in the envisaged future; indeed, a common fault is single-mindedness in vision, and preoccupation with one or two technical or social gimmicks, as if life in the future would not be complex and varied. There is the common tendency, too, of making utopias in one's own image or of envisaging worlds scarcely habitable by imperfect humans. Against these limitations must be set important advantages. Unlike exploratory forecasts of what will be, and planning to avoid the worst, the vision of a 'best' suggests new goals for human knowledge and skills and at least provides landmarks for steering society, a basis for debate.

Most countries of the world are still busy enough trying to improve the material lot of the people and to achieve a high level of affluence and social welfare. This condition has already been achieved in one country—Sweden. It is therefore interesting to inquire where the Swedes think they may be going next. In the mid-1960s there was a lack of direction; a country which, since the 1930s, had led the world in social welfare and the egalitarian ideal seemed about to settle down to become, in the words of the Stockholm architect Ralph Erskine, a 'common or garden affluent society'. Then there were new stirrings in Sweden, largely mediated by an 'ideas debate' in the press and on television. It made many Swedes aware of their relationship to the rest of the world and encouraged the view that the idea of equality and welfare was a poor thing if it were selfishly confined within the borders of one country. Even so, politicians were reluctant to adopt the policies of aid that might curb the rate of growth and the material standard of living in Sweden itself. But pressure mounted for a new international policy. Swedes were rather shocked to discover, for example, that, with one fiftieth of the population, their national income was a quarter of that of India. They also looked beyond their city limits at the natural environment and the problems of pollution, and talked about 'mental slums' persisting in their own country, where physical slums had virtually ceased to exist. The actions of Sweden in the 1970s will be a good indication of whether material affluence and social welfare make people more or less selfish.

Although there is a growing volume of writing about the future, it is mostly concerned with what will be, and may be, rather than with opinions of what ought to be. There is, indeed, a shortage of reasonably cheerful visions. On the contrary, several anti-utopias have gripped the public imagination—black works like Aldous Huxley's *Brave New World*, George Orwell's *1984* and, more recently, Olof Johannesson's *The Great Computer*. Even though the warnings in such books are necessary and important, it may be worse to frighten people out of thinking about the future, than to write too optimistically.

Among the few prescriptive efforts, Elisabeth Mann Borgese (*Ascent of Woman*) took a position in direct antithesis to Huxley's in *Brave New World*; she accepted the artificial breeding and upbringing of children, so vividly and presciently satirized by Huxley, as a fix for the emancipation of women. Mann Borgese looked forward to the day when the sexes would merge, each individual beginning life as a female and finishing up as a male. Any takers?

In the year before his death and thirty years after *Brave New World*, Aldous Huxley published *Island*, a utopia in which a small community of humans discovered the key to an agreeable life in the twentieth century: 'Take twenty sexually satisfied couples and their offspring; add science, intuition and humour in equal quantities; steep in Tantrik Buddhism and simmer indefinitely in an open pan in the open air over a brisk flame of affection.' Any takers?

People are accustomed to be told what they should not do, but are rightly more resentful of instructions about what they should do. Human variation and individual potentiality are far too great to fit unprotestingly in a Procrustean bed of one man's fantasy. Yet something is needed to counteract the purposeless use of science in an age when technology provides its own retrospective goals, by introducing novelties which men then come to regard as indispensable.

That new visions and credible goals are not easy to come by need not be too dismaying; you do not need many of them. One good, ambitious goal could keep men busy for a century or more. But more extensive and systematic attempts to formulate possible positive goals, as a basis for debate, can only do good. Here I can do no more than indicate the sort of visions that I think we need, by some more examples.

1. With the trends towards increased education and increased leisure, we can make amateur study and research, and amateur practice in the arts and sciences, a major element in human life. There are administrative problems of creating the necessary facilities; but more important is the creation of a fruitful interchange between professionals and amateurs, both for teaching and for the practice of useful work by the amateurs. How the status and quality of professional work is to be preserved in such circumstances is definitely a puzzle.

2. The creation of a truly global interchange of information and ideas, based on a computer-communications system, would require much more than technology. Social inventions such as world encyclopaedias, histories and curricula, would be needed, and particularly the development of a world language—either by adoption and simplification of an existing language (probably English) as a universal second language or by the use of an intermediate machine language into which all natural languages can be easily and automatically translated. The latter possibility begs all the difficult philosophical and technical questions at present plaguing the machine translation experts.

3. As a more compactly technological vision, consider individual flight. If men could fly at will, like the birds, independently and safely, that would not only at one blow solve many of the passenger transport problems of the world but would also be a liberation for the individual in his everyday life, and perhaps also a great physical and spiritual stimulus, as Olaf Stapledon envisaged in his *Last and First Men*. It is no longer entirely far-fetched to imagine genetic engineering designed to equip humans with wings, muscles and stabilizing organs necessary to enable them to fly without mechanical assistance. That is the only desirable use of positive eugenics I can think of, but it would be a very long and heroic undertaking, with possible disastrous consequences; it should perhaps, in any case, be forbidden under some general prohibition of genetic engineering applied to men. Alternatively, reasonable hopes of success should now attend a programme intended to develop an extremely compact and reliable personal flying machine—not an aircraft but something more like a combination of lifting parachute and rocket propulsion which would let us fly freely and safely. One obvious problem is of noise, for we should want to be as quiet as the birds, too.

Is 'positive health' a positive goal? It is an old favourite, but René Dubos (1965, see also Chapter 12) argued that we should not deceive ourselves about the perfectibility of the environment and of human health. In 1875 Benjamin Richardson produced a sanitary utopia: Hygeia. It was a city of low density housing, with well-lighted and airy rooms, supplied with scrupulously clean water and excellent sewage and refuse disposal; a medical inspectorate was to enforce the sanitary rules throughout the city, including the factories. Much of Richardson's programme has been achieved in many places, but new degenerative, metabolic or neoplastic diseases have come to take the place of infectious and nutritional diseases. 'There is little to define, recognize or measure the healthy state, let alone the hypothetical condition designated "positive health",' wrote Dubos, who argued that concern for disease rather than health was not simply a mark of bias on the part of the physician, but corresponded to a widespread human trait—health being a disembodied concept which stimulated no emotional response.

A more refined version of 'positive health' is the idea that mental health provides some index of human happiness, and that the goal should be perfect mental health. Whether that is indeed desirable, in view of the creative function of 'divine discontent', is a moot point. As an index of social progress, mental health may anyway prove to be most defective. A Dutch psychiatrist, Robert Giel, surveyed the incidence of mental disease among Ethiopian villagers, whose condition differed in many and radical ways from those of the industrialized Netherlands; yet he found the pattern of mental disease was uncannily similar to that in his own country. Again, if such mental disease should now prove to have a physical or chemical origin, and be susceptible to cure by drugs, that by itself can hardly mean we shall have achieved a perfect civilization. By indiscriminate use of drugs, we may of course kid ourselves that we have done so.

If physicians cannot offer positive goals, who else should try? At the intersection of technology, the social sciences and art, are the architects. They may contend imperfectly with immediate tasks, but they have not failed to think about the man of the future and the kind of environment in which he should live. On the contrary, there is great activity among architects and planners; encouraged by men like Buckminster Fuller they are, in some sectors, in the van of futures research. Visions of new cities can integrate a very great deal of new technological and social opportunities.

Non-scientists and creative artists are badly needed in the constructive commentary upon current civilization and in the debate about what the futures should be. But technology remains one of the main features of our current world and the most powerful instrument for developing a new world; non-scientists should therefore be willing to make fairly careful study, not of the details of research, but of the kinds of applications that can be envisaged. That does not mean all novelists should write science fiction, or painters produce political posters. They should at least be aware of what aspects of human nature are likely to obtrude in the future. How they explore and comment is a matter of artistic judgement.

Changing professional circumstances may stimulate the artists, especially if they dislike the idea of becoming civil servants running an electronic public information service. André Malraux (1967), the novelist and soldier who became de Gaulle's minister of culture, offered this vision:

> 'In fifty years' time culture will be free, for we will be obliged, sooner or later, to treat culture in the way we treat compulsory education today. . . . Despite what some had forecast the machine has not turned us into robots, for the result of mechanization has been that we enjoy unprecedented leisure. Leisure creates its own demand and to meet it we must build factories for ideas just as there exist factories for machinery.
>
> 'This industry for ideas can be either publicly or privately owned. If private then maximization of profit must lead to declining standards; only the state is sufficiently disinterested. Our civilization is under attack. How can it defend itself? In the past we could rely upon religion, today the only possible protection is the intervention of the state.'

Utopia? Anti-utopia? With sharply differing opinions and experiences in various countries of the role of the state in the 'ideas industry', it is at least something worth arguing about, and fighting political battles for or against. That is, in brief, the use for vision.

CHAPTER 16

Who's for a Coup?

WHEN decisions have to be made in the light of expert knowledge and computer print-out, the very idea of popular government is vulnerable. Defences of democracy are makeshift, which merely involve setting expert against expert to curb the power of both. Other political systems seem in little better shape: emperors, colonels and commissars, all are threatened by the march of experts and the demands for modernization.

There are precedents for power struggles between new and old elites. Should scientific experts indeed assume the government of the nations? At a United Nations conference the Pakistani physicist Abdus Salam (1963) was stung by what he called the 'cautious timidity' and 'faltering will' of his scientific colleagues. He reminded them of the enormous powers they had conferred on mankind and rebuked them for turning their backs and leaving the tasks of world development to others. Salam went on to quote Plato's *Republic*:

> 'Until philosophers are kings or the kings and princes of this world have the spirit and power of philosophy... cities will never rest—nay, nor the human race.'

The notion of an experts' take-over is worth more than a frivolous glance, and its hypothetical outcome is instructive. Earlier, we heard Fred Hoyle (1968) putting it to the physicists. He went on:

'Isn't it really that in the past scientists just haven't dared to allow themselves to think openly of taking society by the scruff of the neck? ... Thirty years ago it would have been meaningless to raise the question. We had no experience in large-scale organization. Nowadays this is no longer true. Experience gained in building and controlling large organizations for high-energy physics is already comparable with what is needed to control the economic structure of society.

'My proposals for the 1970s are these: Either keep out of the mess or go right in and change it.'

Every society has its leaders, and whether a particular élite group is compatible with democracy depends on how you or I can join the group, how it operates and what it does. It can gain power democratically only if the qualities the group represents are held high in public esteem. The academic scientists do not have suitable handles of power to achieve a coup unaided. Nor is a grateful democracy likely to confer on them more power than they already have, in the leading scientific nations; Hoyle was probably too late. At the end of the 1960s physics already seemed to have passed its peak in public esteem. The status of academic science as a whole—if it, too, has not gone 'over the top'—may rise to a maximum in the 1970s and then begin to decline as its rate of growth slows down in the leading countries and as social sciences and the arts blossom. The social sciences, which were low in esteem, are already rising rapidly.

Nevertheless, leading liberal-minded research workers may still come to believe that their naïve dictatorship would be preferable to greater calamities. In particular, they might be roused in opposition to a technocracy of government and industrial experts, to challenge them on technical grounds. Or they might throw in their lot with these same experts.

As Kenneth Galbraith (1967) put it, the technostructure is the organization of specialists required for modern technology and planning, who participate in group decision-making in the modern American-style corporation. It is a much larger group than the senior management and extends to the factory floor. The power exercised by such group decision is very great, and difficult for an individual, however senior, to reverse. Galbraith itemizes as the goals of the technostructure: a secure minimum of earnings, which preserves the autonomy of the technostructure; then, the greatest possible rate of corporate growth as measured in sales; and then, as secondary goals, technological virtuosity and rising dividends. The fact that no other social goal is more widely avowed than economic growth suits the technostructure very well, because expansion of output means expansion of the technostructure itself.

The risk of power coming out of industry was identified in America long ago by Alexis de Tocqueville (1838). Of the danger of take-over by a manufacturing aristocracy he wrote: 'If ever a permanent inequality of conditions and aristocracy again penetrate into the world, it may be predicted that this is the path by which they will enter.'

Today, the industrial technostructure makes common cause with government bureaucrats and experts in Eisenhower's feared military-industrial complex and increasingly on the civilian side, too. In a sense, the men concerned do not need a coup; power is already theirs even if the prestige of power is not.

Computers are already immensely important in the everyday running of the most modern states, in both government and industry; during the next two decades they will also take an important place in law enforcement, education and public information systems. The computer experts and systems analysts study and adapt organizations to suit computer operations, and also subject the general conduct of governmental, military and industrial affairs to critical study. They are highly educated, highly intelligent 'ideas-men'; they are young, and taste power daily, but repeatedly they find that older men in authority have only limited understanding of the function and potentialities of computers, and in their work the gap between promise and reality is continual and frustrating.

The computer men have fingers literally on the buttons of power. If they were sufficiently united among themselves, it would be very hard, in a highly computerized state, to keep them under control.

Some politicians already sense the rising power of the computer men. The British minister of technology, Anthony Wedgwood Benn (1967), declared forcibly:

'It's high time that cyberneticists and Ph.D.'s in maths, and programmers and systems analysts, and hardware and software experts, came out of their secret world. They must assume their responsibility, as citizens, for communicating to their very ordinary, ignorant, worried, sceptical and unscientific fellow citizens—like me—and begin assuming citizenship responsibilities for the consequences of what they are doing. They must also understand the implications of all they are planning for us. For it could be the most revolutionary onslaught ever made by any group of men on any established society.'

A coup by the computer men and systems analysts is less likely than a progressive abdication of power to them and their machines. Because of the sheer virtuosity and growing power of computers, computable political problems will be so much more readily dealt with that non-computable problems may be regarded as non-problems. Rather, we men may come to mistrust our own judgement when confronted with the conclusions of a computer. If we set up elaborate computer systems for use in government, and if we programme them carefully and present them with vast amounts of information, it may be very hard for anyone concerned to say that the conclusions should be disregarded because they are not arrived at by human beings.

Already, in military operations, we can see this tendency at work in the computerization of the military intelligence and command structures of the super-powers. It has reached such a point that the headquarters in Moscow and Washington are in possession of far more information about current situations than any local commander ever was before. Calculation and electronics rise like a great tide through the military hierarchies of the world. As ballistic missile defence systems come into play and the response times for technical effectiveness are cut to seconds, it will be extremely difficult to resist further intrusion of the computer into areas hitherto reserved for human decision. International relations are now largely run by the intelligence agencies of the super-powers, as Harvey Wheeler (1968b) has described, and the gathering of intelligence by agents and by spy aircraft, satellites and ships is a continuous war.

All this—including the use of new surveillance systems and computers—is now being matched by domestic police forces, and the technical possibilities are far from being fully exploited yet. Powerfully equipped crime-fighting agencies could easily acquire power similar to that of the secret police of dictatorships. They would do so in the name of fighting criminals or spies.

As the technological strength of broadcasting and telecommunications systems grows—towards the World Box described in Chapter 11—so will the influence of the managers, editors and others concerned in their operation. Already, television is a potent medium, and political censorship of what appears on it can keep a nation misinformed. Proprietors and advertisers can help determine what news and other programmes are provided. The cult of personality that television makes possible may have incalculable effects on political power. Multiply all that by the power of computerized systems and it looks as if the information managers will have to be saints not to abuse their positions. On the other hand, it is hard to imagine them making a successful bid for a special place in the government of the country, unless in alliance with other groups, simply because their activities are inherently conspicuous and transmitters and other equipment are rather easily put out of action.

THE REINVENTION OF DEMOCRACY

Combined operations by some of the groups mentioned could be much more effective than any acting alone. For the sake of argument, one technocratic junta is worth considering.

United action of government planners, the computer men and the academic scientists could create a powerful combination. None of the groups would be over-impressed with the expertise of the others. With all the men in key places having scientific training of some kind, government might be conducted essentially by consensus of experts, strongly oriented towards the future and subject to mutual constraints. If the choice were between total technocracy and total war, or between total technocracy and total inability to govern the onrush of current technology, there is no doubt that we could learn to live under the eggheads.

It might be government of eggheads, by eggheads, for eggheads, with science rated as the supreme human activity and with people of limited intellectual or educational attainment reduced to a kind of affluent serfdom. Yet that is unlikely. More probably, in such a hypothetical fit of power, the eggheads would, on the contrary, come to subordinate their policies to the wishes of the public. If they were reasonable men, they—and particularly the academic scientists—would be struck by the limited role of reason and computation in human affairs, and the methodological impasse of the forecasters.

The areas of human activity in which rational, efficient action is deemed appropriate grow yearly. That is fitting, because knowledge, understanding and power grow. Men no longer beat a stupid child to make him learn, nor sacrifice goats to make rain, nor neglect to use computers to identify stolen cars. It would be equally irrational—a denial of what physiologists and psychologists both can and cannot tell about human nature—to suppose that unmodified reason and calculation can govern men satisfactorily. They need as much rationality as they can get, but they must not, for that reason, neglect the emotions and excitements of life, or the subtleties of human intercourse, or the politically crucial areas of emotional commitment.

As the neuro-scientists would tell their fellow conspirators, all anyone can yet say with confidence is that the human brain operates on quite different principles from the computer. It is an electrical machine, with waves of electrical activity sweeping through it, but it is also a chemical machine. The brain cells depend upon chemical processes and memory involves, at least in part, the production of large and complex molecules. The brain, unlike a computer, is an integral part of that body, sharing in all its mechanisms and experiences. From birth, the 'programmes' and 'data store' accumulate slowly in the context of intense emotions. Fear, anger, sorrow and so on serve definite biological purposes and are not inherently contemptible.

Few biologists are likely to subscribe to Arthur Koestler's (1967) view that evolution erred and left us with a 'new' superior brain superimposed on the old primitive one (the seat of the emotions) 'without providing the new with a clear-cut, hierarchic control over the old—thus inviting confusion and conflict'. Koestler wanted a drug whereby the hierarchic order was restored. When the rulers took his pill, Koestler said, 'then, and only then, would the world be ripe for a global disarmament conference'.

If reason and the upper brain are wise and emotion and the lower brain are foolish, an easier course than seeking a chemical means to hierarchical control would be to abdicate government of our affairs to the computer. Koestler's idea, like the expectation of machines more intelligent than men, requires us to look again at the roles of reason and emotion in political affairs. A little reflection shows that reason can serve good and bad, wise and foolish purposes, indifferently. Emotional commitment to good courses of action is a necessary preliminary to wise reasoning or computation.

Which leads us back to the impasse of the forecasters. Their forecast depends upon what decisions will be made by human beings, now and in the future. They cannot tell even reasonably reliably what those decisions will be without at least asking the people who will make the decisions; yet those people will want to know the forecast before they say what they would do. This logical circle of impotency can only be broken by dialogue. Nor are the decision-makers only high-level people in government—everyone, whether as producer, consumer, student or political activist, makes decisions. Short of extremes of coercion, the system of futures planning can only work by continuous and far-reaching consultation.

Where wealth is widespread and men are capable of diverse feats of technology, choices are so wide open that a primitive situation arises in which every man's opinion is worth as much as any other's, regardless of what he may contribute to the community as a whole. If the general purpose set for knowledge and skills is to make life better for all individuals then it is for those individuals to decide what they mean by 'better'. Given automatic systems for doing essential work, a new order of democracy and freedom becomes possible, despite the fact that modern society is a highly complex and rather delicate apparatus whose overall function has to be safeguarded.

As rational men confronted with a wide choice of technological and economic options, our balanced junta of eggheads should see no virtue in deciding arbitrarily what to do and what not to do. The scientific experts would have no world view to impose. Instead they would wish to inform the public about opportunities, dangers and alternatives for the future, and let the public say in what directions its ambitions lay.

This notion, that the eggheads might have to re-invent democracy in such a fit of power, provides a good starting point for considering the rationale and character of democracy in the scientific age. The acquisition of supreme power may be hypothetical, but the re-invention of democracy is actual.

An international group of experts met to discuss long-range forecasting and planning at an OECD (1968) symposium at Bellagio in Italy. They were moved to issue a unanimous warning 'that social and technological developments already clearly foreseen can exacerbate matters beyond any hope of peaceful relief'. In outlining the remedy that humane planning might offer, the Bellagio Declaration noted: 'Planning must be understood in relation to the consequences and in particular the consequence to the individual of decisions and actions within social systems. It should therefore be performed at the lowest effective level to make possible a maximum of participation in the planning itself and in its implementation.'

As Andrew Shonfield (1965) pointed out, the experts of the French Commissariat du Plan, on their own initiative, became particularly insistent on more publicity, more democratic discussion and more parliamentary control over the Plan. The National Assembly was asked to decide, for example, early in the preparation of the Fifth Plan (1966–1970), whether economic policy should emphasize increased wealth or more leisure; at a later stage, the parliamentarians were invited to consider the detailed implications of the broad choices already made. This initiative of the planners was in fact contrary to the apparent wish of the National Assembly, the members of which had made no serious attempt to exercise control over the Fourth Plan. Shonfield commented that the French planners were by then committed to the creation of the just society, not merely the efficient one which had engaged them after the second World War.

To the even more striking re-invention of democracy in Prague, we shall attend in the next chapter. Meanwhile, having speculated about counter-democratic exercises of power by the expert, we should see where expertise can function most benignly.

TECHNOCRACY SET FREE

Alexander King calls himself an intellectual entrepreneur. He is a burly Scottish chemist who, from his office across the road from the Château de la Muette in Paris, presides over the scientific affairs of the Organization for Economic Co-operation and Development (OECD). The OECD is the rich nation's club with some poor members like Greece—an association of governments of Europe, North America and Japan, with various functions of mutual benefit to the member nations. King has built up his science directorate as the biggest and most active group within the OECD.

King was in at the origins of modern science policymaking during the second World War, as a protegé of Henry Tizard. His chief coup was in spotting the Swiss development of DDT, which the Germans missed and which gave a profound medical advantage to the Allied forces. He went on to run the British scientific mission to Washington and was therefore able to observe the rise of science both in Britain and, under Vannevar Bush, in the USA. After the war he worked with Herbert Morrison, who was at that time responsible for civil science policy. But King came to learn, painfully, that administrative enterprise was not always appreciated in government. At the OECD he found he was able to work without inhibition from the niceties of national policy—to undertake studies, to catalyze international exchanges and generally to generate ideas which the member nations of the organization might adopt or ignore as they chose.

In this role King was a prototype international technocrat, among the first of a species that will assume an important political function in the coming decades. Beyond the administration of his own quite small staff and his visiting consultants and research fellows, he has no power whatever. But he has enormous long-term influence on the activities of governments and industries in much of the Western world; his work thus affects the deployment of big resources.

As 'intellectual entrepreneur' he brought together the ministers responsible for research in the OECD countries and provided them with a workshop in which they could sharpen the ideas and tools of science policy. International comparisons and nation-by-nation investigations of arrangements for science and technology exposed, not merely to public but to international gaze, sensitive areas of government policy.

From a more general point of view, the fact that King's work is in part concerned with science policy itself is quite incidental; he might, in other circumstances, be looking from an international vantage point at insecticides or air transport or nutrition —as, indeed, other international technocrats are doing, in agencies of the United Nations and elsewhere. The important thing is the form, rather than any specific content: the existence of expert individuals or groups, without national loyalties or responsibilities, who are able to stir the pot of ideas at governmental level, to draw attention to common problems, and to throw up innovations of their own. They do not work in the abstract for they deal with real countries, living experts and awkward politicians. They are ultimately subject, of course, to a broad approval of national governments comprising their organizations, but they have a quasi-academic immunity.

The international technocrat provides, in principle, a near-perfect solution to the classical constitutional problem: how to reconcile the power and influence of an indispensable but rather esoteric elite of experts with the rights of public and parliamentarians in democratic government. He works under international scrutiny, so there is little chance of his involvement in secret decision-making. He can make recommendations for action; but a national government is free to choose to accept or reject his advice. It will be a decision visible to the public, and therefore subject to debate; at the same time the national government acts as buffer between its people and the prescriptions of the international technocrat.

Apart from this constitutional merit, the international technocrat has a number of important practical advantages, including the following:

His decoupling from national politics allows him to escape pressure from vested interests of groups or governments.

He is economical, in the sense that his analyses and many of his prescriptions will tend to have wide utility.

He will automatically tend to reconcile the policies of different governments, and thereby reduce conflicts and frustrations.

He can take governments at their word in their pious declarations (about world development, international co-operation, pollution control, and so on), surprise them by showing how action can proceed and then shame them into taking action.

Many of the big issues in technology (supersonic flight, ocean engineering, fuel and power supplies, and so on) have a strongly international flavour and can only be approached seriously from the basis of multi-national experience and opinion.

In the long run the international technocrat cannot escape some ideological or administrative confrontations with national governments. Sometimes he is victorious; sometimes he is defeated, in conformity with the constitutional principle that keeps him in check. A spectacular defeat was suffered by the director-general of the World Health Organization (WHO, Geneva) and his fellow medicocrats when their ambitious scheme for a World Health Research Centre was thrown out by representatives of national governments in 1965. The central idea was for a big international animal-experimentation laboratory—a 'megamouse project'. In it drugs and many other chemical agents newly introduced into the human environment would have been screened for side-effects, including cancer induction and genetic and congenital damage to unborn children. Basic research would have sought to establish general actions of chemicals on humans and other organisms. At about the same time as this proposal was rejected, the same staff of the WHO gained a striking victory in winning approval for field programmes associated with birth control, as part of their technical assistance to less-developed countries, in the teeth of opposition from the Catholic countries.

Although, financially, these birth-control activities were trivial compared with the $140-million budget envisaged for the World Health Research Centre, politically the decisions were of comparable importance. In a hidden way they were irreconcilable. Once the involvement of WHO in birth control was established, medical questions arose. Major technical advances with the Pill and the Loop opened up exciting possibilities for world-wide family planning. But no one could tell what might be the long-term side-effects of the administration of hormones which worked by deluding the woman's endocrine system. Indeed, after a while it turned out that the Pill seemed slightly to increase the risk of thrombosis in women taking it. On the other hand, nobody was sure how the Loop worked, let alone what effects it might have. So what were the risks versus benefits of such materials, used internationally and in very large populations of patients? That was precisely the kind of problem that the projected World Health Research Centre should have been investigating. The governments empowered the WHO to involve itself in programmes using these and other materials, in what would be the biggest public-health operation in history, affecting hundreds of millions of women. They denied it experimental facilities to check the techniques.

In December 1967, the UN General Assembly made a rare intervention in the research policies of member nations. A resolution (2319 XXII) requested governments to report to U Thant, within six months, what they were doing and proposed to do in the way of projects aimed at increasing the supply of protein for human consumption. It resulted from the work of that group of part-time international technocrats, the UN advisory committee on the application of science and technology to development, encountered in Chapter 13. They had already recommended to the Economic and Social Council a world-wide plan of action for averting the impending protein crisis in the poor countries, based on a study by Nevin Scrimshaw of the Massachusetts Institute of Technology. Besides outlining measures to increase protein production from conventional crop plants, livestock and fisheries, and for reducing waste of protein foods, their specific proposals included:

acceleration of the development and growing of genetically improved plants of high protein value;

expanding the use of oil-seed meals as direct sources of protein in human diets;

supporting the production and marketing of acceptable fish-protein concentrates for human consumption;

a great intensification of research on single-cell protein sources, which would be independent of agricultural land;

supporting the use of synthetic amino acids or protein concentrates as supplements to cereal and other plant proteins;

supporting the promotion and distribution of suitable protein foods in the developing countries.

Such was the categorical list of technological projects on which the General Assembly urged governments to make constructive comments and report research in progress. At the very least it obliged governments to assess their efforts in what the international technocrats had pinpointed as urgent tasks; the optimists hoped research would be accelerated significantly and the day brought nearer when, for example, yeasts, bacteria and other micro-organisms ('single-cell protein sources') might help to feed a hungry world.

Given this concept of the theoretically powerless, but highly influential, international technocrat, the creative role of research and technology in human affairs can be enhanced without endangering anyone's democratic or national rights. The existing international science offices—regional and world-wide—were set up in ad lib fashion and they vary greatly in quality and effectiveness. There are advantages in an untidy plurality of agencies, but it will be desirable to strengthen the planetary agencies, especially the multi-disciplinary ones. The specialized technical agencies of the UN show mutual jealousy in areas where they overlap, which dismays even the most sympathetic outsider and does not encourage wise integration of development activities.

Particularly deserving reinforcement are the operations of the UN director for science and the UN advisory committee on the application of science and technology to development, in their efforts to indicate priorities and to promote research in the rich countries. Recall also the effective US mission that investigated waterlogging in West Pakistan (Chapter 8); operational research and systems analysis made available as a continuous international service could assist planners in all the poor countries. Technological forecasting is another area for which international activity is very appropriate, not only to pool and contrast forecasts from all sides but also to give world scientific leaders a basis for relating current research to human goals. As for the unintended side-effects of technological change, the scheme for a World Health Research Centre might be revived with a broader remit: to collate experience and warn of dangers attending all new technological activities.

New international science groups of such kinds would depend, like the much bigger intergovernmental technocratic operations that already exist, upon the goodwill of national governments. Peter Menke-Glückert (1967), a science administrator in Bonn, called for the creation in each country of departments for 'peace planning' with cabinet status which would be based on existing agencies for disarmament and overseas development and whose ministers would have the power to put money into development of 'peaceful future technologies which are designed for international co-operation'. Menke-Glückert looked to new structures of science and new concepts of science policy applied to strengthening international agencies, economic planning and future technologies—to create a new political technology.

Already, operations like those of the World Weather Watch, the International Telecommunications Union and UNESCO in its oceanographic role, give the international technocrats admission to 'hot' areas of advanced research and technology. The International Atomic Energy Agency found it necessary to set up its own nuclear research laboratory at Seibersdorf; as a residue from its abortive project for a World Health Research Centre, the WHO secured a small epidemiological research unit. Scientific research in support of the international activities themselves is likely, by such precedents, to become in time an important feature of the world scene. If, for example, a UN agency were to become manager of the oceans on behalf of the world (see Chapter 10) it is hard to see how it would hold its own against boisterous nations without some research ships of its own.

Whether international collaboration in research should be promoted for its own sake, as a means of promoting international goodwill, or judged strictly in terms of efficiency and utility, is still a matter of controversy. Pugwash scientists, at a conference in 1961, dreamed of setting up an Intercontinental Science Centre in Berlin with $5000 million invested in a 'globular cluster' of institutes, machines and equipment concerned with high-energy physics (with a 300-GeV machine), controlled fusion research, creation of new heavy elements, molecular biology, world health research, and development of giant computers. 'It is our belief,' they said, 'that the astute location of such a striking epitome of science—the most characteristic theme of our modern civilization—could have extraordinarily great significance in improving the tone of the present political situation.' In practice, though the European physicists pursued the idea of a 'world machine' for high-energy physics, the Americans and Russians did not respond constructively at that stage. But the next generation of high-energy physics machine (1000-GeV) may indeed be a world enterprise.

A case can also be made for an International Science Foundation for the support of a part of research around the world.

Eugene Skolnikoff (1967) saw in the international sponsorship of science, not the elimination of personal misuse of the new knowledge, but the presumption that international means would be appropriate to its control. It would also help to reduce the fear of scientific or technological surprise. 'Without doubt, the world will face the question of control or suppression of technology increasingly in the future, perhaps with regard to developments even more frightening than nuclear weapons in their power to influence the global environment or human heredity. And it may well be time to consider more seriously the internationalization of science support as a means of improving the prospects for control of dangerous technology.'

If a really massive increase occurs in capital aid and technical assistance to the poor countries, on a multilateral basis, existing UN agencies may be unable to cope, without bureaucratic over-growth. Decentralized modes of working are, however, foreseeable, in which national governments provide offices and staff to international planning and operations, as human efforts swing in the direction of world development. Better still, scientists, planners and other experts will be recruited and deployed in universities and industry, working for the international agencies but dispersed in an Invisible College all over the world.

The huge corporations with world-wide operations, like General Motors, Shell and IBM, will meanwhile be growing in scale and influence. They will probably also change in character, to meet criticisms or outright revolts against commercial colonialism. Their activities, like those of the international technocrats, will be increasingly concerned with planning and with the international transfer of technology. As Richard Robinson (1967) put it, the international corporation of the future will probably sell its technologies, its skills, and its distribution services on a contractual basis to overseas firms that are largely locally owned. 'Under these circumstances, private international business will probably generate a net pressure in the direction of closing the gap between rich and poor nations. If so, its life expectancy is vastly increased.'

The international organizations, companies and planners will gradually become altogether too influential for the liking of governments. More and more of the national budgets will be spent in multinational agencies under the guidance of the international technocrats. World political institutions will have to be strengthened, to supervise these agencies and civil servants. That is how world government will develop—not in an idealistic leap, but by accretion of activities and institutions, many of which already exist, made natural and necessary by global policies and technologies. Nations will find their purposes for existence emptied into the international pool. Even education may pass to international universities and teaching programmes. The international technocrat may then succeed, where prophets of hellfire and preachers of the brotherhood of man have failed, in uniting the human species.

CHAPTER 17

Democracy of the Second Kind

THE DEMOCRATIC revolution in Prague in 1968 blasted the residual Stalinists out of office as thoroughly as some years previously the Czechs had dynamited the huge statue of Stalin himself that had towered over the capital like a Golem. The revolution which Czechoslovakia's Soviet allies found 'absolutely unacceptable' came as no surprise to those who had followed currents of thought in Prague in the mid-1960s. Like the Vltava river that runs below the castle where Tycho and Kepler worked for Rudolph, and goes into the Elbe and the North Sea, there was an apparent westward flow; but to see the process as a kind of subversion by the anti-communists of the West would be to accept Moscow's version. To a great extent, the scientists and their academic colleagues were responsible for the build-up to the revolution, and not only for the elementary reason that good research and dogmatic ideas were incompatible. The technocrats had to re-invent democracy to make sense of their administration and their goals.

At the time of the Soviet invasion of Czechoslovakia, I noted of the 'Czechnocrats': 'We all have a lot to learn from the Czechs. Starting from a higher political level, they have thought more deeply than any other nation about the impact of current science and technology in a modern form.' (Calder, 1968b)

Even the Novotny regime in its last years was highly science-minded. The idea that political leaders should take account of technological forecasts in developing their policies had, in Czechoslovakia, ceased to be a matter for pious pipe-dreaming. It became a part of routine administration. The president of the Academy of Sciences, František Sörm, an authority on the chemistry of plant oils, and politically conservative until the Dubcek period, held ministerial rank in the governments as *de facto* minister of science. Ten of his Academy's specialists, along with technologists from industry, were members of the central committee of the Party.

The Academy dealt with 'pure' and long-term applied research in its own one hundred and forty institutes and laboratories, with the universities, and also with medical research. On the Soviet model, it was paralleled, after 1962, by a state committee concerned with research and development in industry.

The Academy had an Institute of Science Planning, also set up in 1962, where a score of graduates, half of them specialists and half generalists, co-ordinated ideas about the course of science up to about 1980. The directors of the Academy's institutes and its scientific boards were required to look both at the trends in world science and possibilities and needs in Czechoslovakia over that period. On that basis, the Institute would draw up the general five-year plan for discussion and compromise within the praesidium of the Academy. The plan also went to the government for particular consideration by the state committees for planning and for science and technology, and for integration with plans of the Ministry of Education. Conversely, all the long-term plans of the government were considered by the scientists. The declared aim was to match 'the suppositions of politics and the visions of science'. And as one member of the staff of the science and planning institute put it to me, 'Quite new things are surprising, but not unexpected'.

That was not perhaps sufficient proof of ultra-modern political ideas. Systematic forecasting of the scientific trends, looking twenty years ahead, occurred elsewhere, including other communist countries. What was rare was the re-examination of inherited political teaching, by scientists and politicians together. From about 1965 onwards it was openly subject to revision in quite drastic ways, and the belief that Marxism was a science-based political philosophy was interpreted to mean that great weight should attach to the voice and opinion of men of learning.

By the continental meaning of 'science', the Academy of

Sciences included the social and political sciences; and the Academy's economists, under Ota Sik, were primarily responsible for the politically sensational new system of economic planning. That began to take effect in 1966, and it provided for greater independence and initiative among industrial managers, for the unfreezing of prices for goods, and for 'material interest' of individuals in the success of state enterprises. At the time of the Soviet invasion of Czechoslovakia, Sik was deputy prime minister.

In Prague, not long before the Dubcek revolution, I talked with Jaroslav Kozešnik, vice-president of the Academy of Sciences. Formerly chief mathematician at the Skoda Works in Pilsen, he was at the time director of the Institute of Information Theory and Automation, and a member of the State Planning Commission; an intense but cheerful man of the kind common among the higher technical echelons of big American enterprises. In the Academy he was in overall charge of technical sciences, and also responsible for science planning. For a small Marxist state the slogan 'Science is a productive force' required re-examination, Kozešnik said, if only because the links between the specific fundamental research undertaken in the country and applications of science in industry were necessarily tenuous. Czechoslovakia contributed only about one per cent of the world's fundamental science, while its industries manufactured perhaps 70 per cent of the classes of all products made in the world.

So what was the role of research in a small country like Czechoslovakia. Kozešnik saw a purpose in establishing 'the moral right to use the world results'. More practically, without good indigenous science the Czechs would not understand or evaluate the 'world results' that might be useful to them. Most significantly, fundamental science was necessary for creating an intellectual climate in which scientific ideas were readily entertained. That was the new gloss that the Czechs were putting on Marxist ideas, recognizing that the relation between science and everyday life was much less idealized and much more complicated than orthodoxy would have it.

Also seminal in Czechoslovakia in the mid-1960s was the Academy's philosophical 'futures research', headed by Radovan Richta (1966 and 1967) and Ota Klein (1967) of the Academy's Institute of Philosophy. A 45-man team of experts in many fields undertook broad and forward-looking studies on 'the character and context of the scientific and technical revolution'. They discussed the effects of technical change on the style of life, on the working milieu and management, on leisure and education, on health and on politics. Aspects of the technical revolution that particularly struck Richta and his colleagues included:

the abolition by automation of the split between management and workers, and of the 'intelligentsia' as a separate stream of society;

the need for education to be continued throughout life and the opportunity for 'individualizing' education with the aid of teaching machines;

the need to fill growing leisure time with creative cultural activities;

the menace of 'technical one-sidedness', and its remedy by 'aesthetically strong and emotionally appealing art';

the risk to mental health of rapid changes in the process and rate of living; these changes needed therefore to be directed towards the 'humanization of man';

the problem of technocracy versus democracy;

the requirement for participation by all workers in discussions about the long-term perspectives for the whole society.

With automation, the major part of human work would take on the creative character of 'active self-assertion' and the educational level of the working people would have to be raised to that of the contemporary intelligentsia. In the view of these scholars, mindless expansion of material production would soon lead to irrational economic waste; better returns from society would come from expenditures on welfare, education, increased leisure, improved working and living conditions, greater social security and the like. 'At a certain level of growth the most effective method of expanding society's productive forces is inevitably found to be the development of man as an end in itself.'

Richta and his colleagues sought 'civilization regulators'. For the social control of the technocrats they offered, as 'the only definitive solution', that of making all workers 'socialist specialists' by wider education, by changes in work, and by 'various forms of participation in civilization'. That could not be achieved at once—workers' participation in the scientific and technical revolution would be neither immediate nor automatic. But the philosophers looked to an increase of conflicts of ideas in a context of 'new forms and new rules differing from the ideological struggle that accompanied class actions', seen as a dialogue in which there would be no distinction between informed and uninformed parties. 'Where the social value of the individual development of all people increases, individual development should be honoured and respected as a collective value... the new position of the individual in collective struggles of the scientific and technical revolution will finally be the most conspicuous change in the character of the historical process.'

Amid the Marxist jargon the alert reader of this idea for a participating democracy can discern, besides decentralization and reward for initiative, demands for free speech appropriate to a scientific age, and also the general proposition that the citizen of the future, hopefully well educated, is entitled to participate in the decisions that shape his own future. As Richta's group put it, it was necessary 'to transform the creation of perspectives into a public long-term matter concerning the whole society, organized by means of continual discussions about various alternatives and possibilities with a large participation of scientists, technicians, experts and all workers.... This method of forming the perspectives corresponds to the conditions of the scientific and technical revolution as well as to claims on an informal participation (especially of youth) in shaping one's own future' (Richta, 1966).

The new approach to democracy in Prague, which began to 'take' politically in the first half of 1968, was more than simple nostalgia for old-fashioned Western styles of parliamentary democracy. As to the political status of the studies of Richta and his colleagues, we have the testimony of Sörm (1967), the Academy president:

> 'The project was discussed in the Academy of Sciences and at top political levels. It was ultimately decided to utilize it in developing the theoretical and practical aspects of social advance and in preparing long-term guidelines for a country which, having set out on the difficult and in many respects untrodden path of building socialism and communism, is searching for new, humanist variants of a technologically advanced civilization.'

The arrival of the Warsaw Pact tanks in August 1968 not only threatened the lives and freedom of Czech citizens. It demonstrated the resistance of Eastern communist orthodoxy to the most elementary social developments needed in a science-based state; more than that, it smothered, albeit only locally and temporarily, the brightest flame of new thinking to be found anywhere in Technopolis in the 1960s.

INVENTIONS FOR DEMOCRACY

In a discussion in Washington, a man who had worried long and hard about the problems of government in the technological age gave a reasoned presentation of the need for machinery for integrating action, for reconciling many simultaneous goals, and for exploiting all relevant knowledge and skills in the community. He then came out with the idea of a National Administrative Research Agency. This would consist of a National Institute for Unified Studies, a Management Sciences Commission and an Operations Research Service. After the speaker had described at length the functions of this agency in evolving policies, a shrewd zoologist present muttered: 'He's just invented the federal government!'

Governments exist and function, however imperfectly, and it would be naïve to expect them to recognize their obsolescence and quit. To add to existing forms of democratic and other governmental control will be easier than to dislodge or amend existing forms. It may be a pleasantry to say, with Buckminster Fuller (1967), that 'politics will become obsolete'. But so long as men live in groups and have to develop common policies there will be the endless interplay of opinions and loyalties, controversies and machinations that is meant by politics.

A six-point outline of a participating democracy was sketched by a British cabinet minister, commenting on the May 1968 revolution of the French students. Anthony Wedgwood Benn (1968c) said: 'We are moving rapidly towards a situation where the pressure for the redistribution of political power will have to be faced as a major political issue.' In summary, his six points were:

1. Public information about government activities. 'The searchlight of publicity shone on the decision-making processes of government would be the best thing that could possibly happen.'

2. Government information about the community, exploiting computers. 'This information could and should compel government to take account of every single individual in the development of its policy.'

3. Participation by the electorate in decision-making. 'Electronic referenda will be feasible within a generation.'

4. Outlet for minority opinions in mass communications. 'What broadcasting now lacks is any equivalent to the publishing function.'

5. Cultivation of representative organizations of all kinds. 'The more representative and professional pressure groups could be, the more government could work with them and power be redistributed.'

6. Devolution of responsibility to regions and localities. 'It means identifying those decisions which ought to be taken in and by an area most affected by those decisions.'

All these points were closely relevant to the renewal of democracy, and the minister of technology showed a good grasp of the political implications of computers and communications.

An 'electronic loop' between rulers and citizens is the most significant possibility for the machinery of government in the foreseeable future. It is hard to mock, without denying democratic principles; hard to ignore because it is already partially in existence and can very rapidly be completed. When Wedgwood Benn looked to electronic referenda 'within a generation' he was speaking conservatively, for any country already possessing an extensive telephone service. Yet, even without those instant polls envisaged by Vladimir Zworykin (1965), the television inventor, existing facilities and institutions provide the means for launching what I shall call 'democracy of the second kind'.

Radio and television are obvious media for publicizing political issues; increasingly they are airing public responses. Television, with well-produced programmes, has advantages for didactic purposes. But attractions of radio are that it will be reasonably easy to dedicate a channel wholly to public debate and that the simplest existing way of recording expressions of opinion from the public is by telephone. Such messages can be readily broadcast, to sustain the debate (see page 340). Polls taken after a particular theme has been debated provide a minimal technique for registering views numerically.

As a means of obtaining information and opinions from the periphery, computer networks can promote decentralization of political power. The technology of such democracy will be greatly aided by advanced computing techniques and systems analysis. As Harvey Wheeler (1968a) of the Center for the Study of Democratic Institutions, California, wrote:

> 'The system would make it possible to refer non-trivial, controversial issues not merely back to lower echelons but back to the people themselves. For even when data processing cannot reduce the choices to one, it is still able to spell out the larger implications in the various conflicting possibilities. For example, suppose the people, if asked, chose a two per cent rate of economic growth rather than six per cent? There is no way for most of them today to make an informed choice,

but information-processing systems could pre-rationalize the choices and so compensate for the technical deficiencies of the man of practical wisdom.'

There is room for corresponding social inventions, too, especially in the registering of opinion, whether by survey or voting. There may be no need to make special provision for the voice of the expert to be heard; respect for his knowledge should ensure that his opinion will be sought. But the activities of pressure groups, or the probings of detailed opinion surveys, can register the strength of conviction of individuals or their level of concern for particular issues. So why not give a man ten votes on ten issues, and let him, if he chooses, cast all ten votes on one that especially exercises him?

Exceptionally important is the new educational process, the growth of which will be greatly aided and accelerated by electronic means. While the open 'university of the air' will become the main pattern for organization of study, the 'teach-in' can become an important mode of teaching in most if not all subjects. The 'teach-in' was invented in the United States in 1965, as a form of political action appropriate to an institution of learning: a striking blend of didactic, dialectical and demonstrative activity.

Add to that the expectation that education will be a life-long business of learning and re-learning—both because of individual and social needs to 'keep up' with change and because of increasing leisure and opportunity for learning. Then you have the higher educational system as the eventual vehicle for democracy of the second kind. It can be reconciled with student power if the teachers will exert their minds and consciences to recover leadership of their own students. It becomes progressively more democratic, in the sense of equal rights, as more people continue education into adulthood; that it will nevertheless retain a built-in bias in favour of the ideas and opinions of youth and to the exclusion of the elderly is a positive virtue.

Just as there has to be an intellectually revolutionary leadership to heave the poor countries out of their poverty, so there has to be a revolutionary leadership in the rich countries to make them adapt their old-fashioned goals to rapidly changing technological and social circumstances. The finger points unyieldingly at the young university teachers of science, technology, social science and the arts, who are already set by society in a position to lead but are young enough to think afresh and to carry the students with them. I am certainly not the first to cast them in that role, and Kenneth Galbraith (1967) looked to the 'educational and scientific estate' to provide the counterpoise and political leadership for the development of fresh foreign policy and domestic policy. He wanted it to reaffirm the complexities of life and the need for much broader goals than those of the industrial system: 'If other goals are strongly asserted, the industrial system will fall into its place as a detached and autonomous arm of the state, but responsive to the larger purposes of society.'

Of course, there will be opposition to such a trend. Edward Shils (1956) said of the USA: 'Although this country owes its creation as a state to intellectuals, there has been, ever since the Jacksonian revolution, a distrust of intellectuals in politics and a distrust of politics among intellectuals.'

But with the university of the near future the distrustful will have to endure politically active intellectuals or do without the university. For that institution will be, not a court of polite study, nor a glory-hole for complacent professors, nor even a fortress for dissident students: it will be the parliament of renewed democracy.

It needs, however, to mesh with existing government and 'democracy of the first kind', also with consideration of the uses of science and with serious discussion of 'exploratory' and 'normative' futures—which brings us back to another theme.

THE FUTURES DEBATE

If a minister decides to start a new aircraft, or a new town, or an extensive agricultural reform, another minister—perhaps a series of ministers—will usually have to see the project through to completion. The time required for many innovations is short compared with the human-life span but longer than the term of a parliament or ministerial tenure. Each administration takes over the policies of its predecessor, in the form of ongoing projects, even if its political colour is quite different.

In 1964, an incoming Labour government in Britain had great doubts about the value of the Concorde supersonic airliner, the Europa satellite launcher, and the Polaris submarine programme, all inherited from its Conservative predecessors. Hypothetically, it would not have started them itself. In practice it felt obliged to continue with them all, although it did cancel some other projects. Time is on the side of the bureaucrat, because he endures while ministers come and go. Nearly all a government's budget goes in honouring previous commitments, so that only a small fraction is available to start new projects.

Political credit can be gained from starting promising new projects, but even more from their completion. Incoming governments are quick to blame their predecessors for current difficulties, but are glad to bathe in the glory of completed motorways or missiles, telephone systems or power stations, put in hand years previously. There is scope for snatching a dazzling project out of the air, just before an election, without actually committing any resources to it. Otherwise, little electoral advantage comes from taking a long-term view. A politician's interest in the future can only be moderated by the little likelihood that he himself will be in charge in, say, fifteen years' time. The situation seems also to diminish the voter's power. Prior commitments mean that changes concerning big projects will usually be marginal, in the short-run, whichever way he votes.

A dictator with a good bodyguard is more durable and long-term planning comes more naturally. The USSR, for example, followed a reasonably consistent policy of development throughout Stalin's reign. But it incorporated errors and crimes on a terrifying scale. Single-minded planning is less likely to pay off, even than the short-term opportunism of labile governments.

Politicians in democracies have the contrary fault; they are obsessed with things that seem important on short time scales corresponding roughly speaking to the interval between elections. The great issues of the present era are on a longer time scale and typically concern the events of one or two decades ahead. It is for this reason, if for no other, that the content and consequences of scientific research seem a matter of relatively little political interest. Yet, as Erich Jantsch (1967) has pointed out, if technological forecasts are coupled with current social goals, there is a mismatch.

The techniques of the planner and forecaster are like methods of navigation: unless you know where you want to go the best sextants and radio aids in the world are useless to you. This is the methodological impasse of futures research encountered earlier, but now to be seen as something met daily by government planners, though they may not recognize it. Whether they are planning a missile system or an infants' school, they have to make assumptions about the political and social context in which the item will operate, during its useful life. They can expect no serious guidance from the politicians. In practice, they will guess for themselves—unless they are merely stupid and assume, unconsciously, that everything else in the world will freeze besides the item under consideration.

In modern industrial and military planning, attention to the future is taken for granted and forecasting is a process scarcely distinguishable from the routine activities of good management or generalship. In the same way, awareness of and study of the future will become an integral part of good civilian government. There is no sign that it will come from the elected leaders. There is, in fact, a power vacuum at the very point where important decisions are taken that determine the social and technological character of the world tomorrow.

The point of attachment to government, for democracy of the second kind, is therefore plainly in the long-range planning groups of the bureaucracy. It is here that the government's own 'future-mindedness' is principally located, but somewhat removed from day-to-day battles and the preoccupations of the political rulers. The long-range planners are looking well beyond the term of the government of the day. Accordingly, they should welcome some non-arbitrary framework of desired developments within which to work, provided by the re-invented democracy.

One then has a two-phase system of democratic government. The elected group in office is primarily concerned with day-to-day issues and with the short-term future; it continues roughly as at present. Another activity grows up—the futures debate—to deal with the longer-term issues and to register opinions about goals twenty years ahead. Again, this is not a pipe-dream.

In 1969, an experiment was put in hand by the Swedish Broadcasting Corporation. On a 'pop' radio channel, aimed at young people, experts speak on opportunities in the future. Listeners can then give their opinions using the telephone system. They can either (1) say a piece, which can be recorded and broadcast very quickly, with editorial control, or (2) vote upon a particular question (dial one number for 'Yes' and another for 'No'). Here is elementary technology for the futures debate and the democracy of the second kind. As another way of registering views, the BBC-TV *Talkback* programme uses a studio audience selected on a statistical basis to reflect all shades of opinion. (Sample question: 'Has science the moral right to start life outside the human body?' 67 per cent said Yes.)

Although opinion-polling will be useful for some general questions, forceful expression of ideas and views against the consensus will be more important. The Swedish broadcasters found it very difficult to reduce questions about the future to a Yes-No form. The main principle should be that everyone's wishes count. Opportunities for leadership are greater than ever, but the expert and the politician must make plans that encourage diversity and individual freedom of choice.

The constitutional relationship of this second activity to conventional government of the day is a matter for adjustment. For a start, research administrators and long-range planners in the government service will be entitled, but not compelled, to work towards goals identified by the democracy of the second kind, unless the government of the day specifically (and publicly) orders otherwise. Industrialists, teachers and research workers outside government are also entitled to use the expression of popular will as their guide.

The chief impact of such articulate public opinion about the future will be in the planning of scientific research and in proposals for technological, industrial and regional development schemes. Decisions will still rest with the normal government and questions of priorities and timetables of action would certainly remain firmly in its hands. From the practical point of view of trying to create the new, co-existent system and make it work, it may indeed be fortunate that science and other long-term issues are still for the most part on the periphery of ordinary politics. This circumstance will make it easier to bring about a quiet political revolution.

Nevertheless, if the debates about the longer-term future are earnest and continuous, the eventual effects on party policies, on the themes of election campaigns and on the decisions of government will be profound indeed. Parties with a weakly developed sense of purpose (which means a poor sense of the future) will be vying with one another to follow the popular will as manifest in the futures debate. Other parties, with stronger purposes, will seek to impose their own visions of the future in the futures debate; if they fail they may, in seeking office, continue to beg to differ.

Allegedly scientific methodologies for forecasting the future should be treated with reserve, but one is of interest because it attempts to gauge opinions of non-experts. At the University of Illinois, Charles Osgood and Stuart Umpleby (1967) adapted a computer teaching system for exploration of the future as a game played by individuals interacting with the machine. The events under discussion were drawn from a game originally produced by the Kaiser Aluminium and Chemical Company, in turn based on a RAND Corporation technique, called Delphi. The 'player' has to try to anticipate future events, to seek interconnections between them (for which a score is given) and to construct 'what he believes to be the best of all possible worlds'. The psychologists concerned developed the system on the basis of twenty simultaneously operating stations and believe that it provides a method of setting large numbers of people to exploring the future and arriving at conclusions about it.

An obvious snag is that the author of the programme has to select areas (twenty-five in the first experiment) for discussion between the player and the machine, in order to keep the programme of finite size. In the preparation of the programme, a questionnaire elicited from faculty members and students in the university some general patterns of belief in the probabilities and desirabilities of particular future events. Results of this preliminary phase of the work were themselves interesting. The events thought most desirable by people responding to the questionnaire were: control of air and water pollution, the elimination of racial barriers, and lifelong education; the least desirable events were all-out nuclear war and US involvement in limited war, though the latter was regarded as quite probable. The most controversial subjects—that is to say those for which there was least agreement about whether they were desirable or not—turned out to be planned population levels, and direct democracy by computerized polling!

THE ABOLITION OF BUREAUCRACY

For convenience of argument the point of attachment to government of the new futures-oriented democratic process was identified as the long-range planning offices of the bureaucracy. But the bureaucracy itself will largely die. The consequent changes in government machinery will assist both in giving experts of all kinds a more vigorous, albeit temporary, part in government and will also bring the futures debate closer to the political rule of the day.

Among the reasons why the total abolition of the permanent central bureaucracy is a desirable, if not quite attainable goal, one is the very threat of a coup by an expert élite within the government machine—or, if not an actual coup, at least a dangerous concentration of technocratic power. In any case, the minister's sources of information and advice will lie increasingly in the outside world, particularly the universities and industry, so that senior bureaucrats will tend to be replaced by consultants and by part-time or short-term advisers. Even when sustained effort over a number of years is required for a particular project or plan, this work can be farmed out to universities or other suitable institutions—perhaps with two or more separate groups at work on the same problem in competition with one another. This tendency is already manifest in many of the scientific functions of government, particularly in the United States.

Such is the likely fate for the creative arm of the bureaucracy, as the source of innovation in the drafting of new laws and the formulation and evaluation of new projects and policies. In principle, a minister may need only a few assistants and secretaries to assist him in his communications with the outside world, and plenty of spare offices for visitors. On the other hand, the routines of bureaucracy such as administering pensions, tax collection and so on, can be computerized.

The elite of a future, futures-minded, democratic society will be located in the universities. They will be the new generalists, students of the principles and implications of the latest knowledge in all areas, with a trained awareness of the gaps in knowledge and of the complexity of human affairs (now and in the future). They will be like neither the 'amateur' political and bureaucratic chiefs of today, nor their technocratic experts. Unlike the first they will be accustomed to sustained study and research and creative thought; unlike the latter, they will not think physics, or economics, or engineering, or any other expertise supremely virtuous or preoccupying. If one were to offer a basic syllabus for the training of these men, it would certainly include mathematics, computing, ecology, literature, history and psychology, but also sufficient of the atomic and molecular sciences to save them from being blinded or baffled by them. Most important of all, they would be educated to penetrate respectfully but fearlessly into any branch of learning or human affairs relevant to the issues of the day.

CONDITIONS FOR CONTROVERSY

The futures debate, using the machinery of the democracy of the second kind, will examine all aspects of social life and draw upon all branches of learning. The many technological issues rehearsed in this book represent only a fraction of the field. Nevertheless, my main concern is with the debate of these uses of science, and in this connection a simple and valuable rule can be stated:

Oppose everything!

Whenever any non-trivial new application of science or technology is proposed, reasonably well-informed people outside the decision-making process should advance all the arguments that they can think of, why it should not be done. This opposition should begin when the idea is just a gleam in somebody's eye, and continue at least until the point when the relevant government, company or international agency has taken an irrevocable decision to proceed.

Support for the idea that, simply as a matter of good administration, someone should play 'devil's advocate' to oppose any project came from no less an authority than James Conant (1952), the former president of Harvard University who was chairman of the wartime National Defense Research Committee in the USA. Talking about the review of possible innovations within the US Defense Department he said:

> 'The important point is that there should be arguments *against* the proposal: they should be vigorous but candid; a technical expert should speak on behalf of the taxpayer against each large proposal. . . . With opposing briefs, arguments, and cross-questioning, many facets of the problem, many prejudices of the witnesses would be brought out into the open.'

In a report to Congress (Brooks, 1967), the National Academy of Sciences warned the politicians that technologists committed to a particular line of effort tended to minimize difficulties and underestimate costs, while research workers were often 'over-conservative' and inclined to call for more research, underestimating the applicability of their science. 'In appraising the situation, it is important for Congress to listen to the sceptics as well as the enthusiasts, and to ask the enthusiasts to answer the arguments of the sceptics. Laymen can learn a great deal from the confrontation of experts even when they do not understand the details. Especially in applied science and technology, priorities and goals can be established only through a multi-dimensional interaction between scientists, technologists, public servants, and the general public.'

The high-level use of computers in administration is a particularly important case for contest from those who doubt the computability of human factors and sense the non-rational root of our value systems. Important applications of computers in government should be challenged routinely and repeatedly, to bring to light and debate the implicit goals and assumptions.

Military and commercial secrecy need not nullify this procedure. At worst, opponents can be found who are privy to the secrets. What is more important, application of radically new discoveries or inventions takes time and, as a leading research director (Speiser, 1963) has testified, general technological ideas tend to be current among all research workers in a field before the 'wraps' go on a specific development. Nor do I see that such opposition need be a brake on enterprise, except where a brake is socially desirable. The debate can run concurrently with technical studies and, where an idea is really good, public discussion may encourage an increase in effort for it, and more rapid acceptance of the innovation when it arrives.

The chief difficulty is that public controversy does not come easily to the scientist, especially if it means dispute with his colleagues. His fears on this score have at least five origins: first, in the habit of striving for consensus in research itself, which he translates, by confusion, to a striving for consensus about the uses of science, which are not 'scientific' questions; secondly, in the feeling that 'scientists must stick together' in the face of latently parsimonious public and government; thirdly, in fear as an individual for his position (or his grant!) if he becomes engaged in public disputes; fourthly, in a wish not to compromise his status in the public eye as an infallible expert witness. None of these reasons is very creditable. Only the fifth deserves some sympathy: a professional reluctance to speculate on technical possibilities that are not yet established. The wish for unanimity produces absurd behaviour: in the USA in the 1960s efforts to strengthen the President's Scientific Advisory Committee with social scientists were resisted, reputedly on the grounds that divergent views on technical subjects would be generated within PSAC.

All too often, experts rise in protest long after a wrong decision has been implemented, and only when sufficient evidence of ill effects has accumulated to make them feel confident. That was just what happened in the case of nuclear weapons tests and the indiscriminate introduction of pesticides; Linus Pauling and Rachel Carson had to stick their necks out first. Experts do sign petitions. They are willing, too, to subscribe to unanimous resolutions at the end of conferences. But they feel dismay when prominent members of their profession contradict one another in public. In fact, of course, there is much controversy within scientific circles, and when these differences of opinion reflect genuine uncertainty, dilemma or ignorance, the public and politicians should know about them.

Research workers do not often take the initiative in laying important new technical possibilities and problems before the public. This situation is improving in the USA in some respects, for example in the work of the Scientists' Institute for Public Information and in general journals edited and written by research workers, notably the *Bulletin of the Atomic Scientists* and *Science*. (As the St Louis group of SIPI explained: 'We don't have to take vows of political chastity here.') More commonly, the alarm bell has to be rung by the science journalist or newspaper editor or television producer, spotting the implications of learned papers or research projects and perhaps inviting the experts to say something publicly. But that is a hard task for under-informed and over-worked publicists; experts who are aware of important breakthroughs or issues of public policy arising in their areas should make it their responsibility to see that the public is told.

There is endless rhubarb among scientists and non-scientists about 'the social responsibilities of scientists'. I believe they come down simply to this:

Speak out!

But will the public understand the scientists if they do speak out? A common yet unnecessary bar to sensible regulation of technology is the widespread belief that modern discoveries and inventions are difficult and so voluminous that nobody, least of all a non-scientist, can master them. Superficially, this is entirely true; no individual can know except in the broadest outline what is going on, and what is already known, in all fields of science and technology. There are masses of detail comprehensible only to the specialist which fill vast libraries of books and periodicals. Nevertheless, the important things, intellectually and politically, are certainly not infinite. Any thoughtful person can share the magnificent picture of the natural world, from atoms and galaxies to living systems, that research has painted; he can grasp the principles of things like nuclear fission and fusion, polymer chemistry, spaceflight, the genetic code and so on; he can understand uses and possible consequences of the principal technologies with which he is surrounded and which are emerging for the future. And he can safely leave the details and know-how to the experts.

Even with the aid of painless popularization of science and technology, by press and television, some time and mental effort are still needed, to take in principles and consequences. But it is not an undertaking of a different order from, say, following fashions in the theatre or the form of racehorses. The chief consolation is that the aim of scientists is not to amass information, though this may be a necessary phase of their work. They want to explain natural phenomena as simply as possible. Progress in research is judged by its success in reducing a complex and untidy agglomeration of information into a concise 'story'. It therefore becomes possible for the scientist, and through him the non-scientist, to know incomparably more about particular subjects, but in succinct form. The tendency is towards broader theories and clearer principles.

Controversy about the uses of science will not *resolve* the differences of view, as a noisy prelude to consensus. On the contrary it will exacerbate differences, by making people sharpen their ideas and adopt debating stances. That is not a bad thing at all. Sustained controversy will ensure that decisions tend to favour diversity and pluralism and to broaden personal options rather than narrow them.

MUGS AND ZEALOTS

If the futures debate is given an appropriate dose of controversy, a new axis of political opinion and confrontation will become important, quite different from the conventional radical-conservative axis. Some hint of it appeared in Chapter 14, in the hypothetical parties Z and M, dedicated to technological opportunism and scientific conservationism. In fact this distinction represents a deep-rooted variation of human types, much more ancient than science and technology.

Psychological studies of political attitudes make the most sense when one compounds the straightforward placing of political attitudes along a conservative-radical scale with the further distinction between tough-minded and tender-minded people (Eysenck, 1954). That leads to a distribution, which British and American studies made very convincing, between tender-minded conservatives (Conservatives), tender-minded radicals (Socialists), tough-minded conservatives (Fascists) and tough-minded radicals (Communists). At a neutral position on the conservative-radical scale, but with a high rating in tender-mindedness, fell the Liberals.

The 'T-factor', as the psychologists call the measure of tough or tender-mindedness, plainly has a strong bearing on the temper of politics. To a much greater extent than the distribution of opinion along the conservative-radical scale, it seems to be closely related to the individual's psychological makeup. Jung divided human personalities into two types, extravert and introvert, terms much abused in everyday usage but meaning, technically, people whose interests and energies are directed outwards (extravert) or inwards (introvert). In the psychiatric ward, the disturbed extravert is likely to suffer hysteria, while the introvert is likely to be an anxious neurotic. From the political point of view the interesting discovery is that extraverts tend to be tough-minded and introverts tender-minded. As a group, the extravert, tough-minded people tend to be submissive to group pressure and to prefer simple ideas, while the tender-minded introverts show greater independence of judgement and a taste for complexity.

The real difference in attitudes to the uses of science lies along that other axis—of tough-minded versus tender-minded attitudes, hysteria versus neurosis in the pathological extremes, technological opportunism versus scientific conservationism in the futures debate. The regrouping is thus more firmly based than it may have seemed at first sight. There is more to it than expedient distinction between competing technologies; it reflects a difference between people more fundamental than the old conservative-radical difference.

The clash of opinions can now be clarified, which characterizes the political struggle about the uses of science that has already begun and will intensify. The tough-minded and the technological opportunists I propose to call Zealots, for self-evident reasons. The tender-minded and the scientific conservationists I call Mugs, this term serving as a shorthand for mugwumps (people who regard themselves as being above party politics) and also having an appropriate slang connotation of passivity.

It is, of course, a great over simplification ever to attempt to divide human ideas, let alone human beings, into two mutually exclusive sets, 'Big-enders' and 'Little-enders'. But the paradigms of political rivalry, of democratic choice, of advocacy and refutation, compel us to conform to this primitive binary logic, even though it favours zealotry. In practice Zealots disagree with one another because their beliefs are incompatible; Mugs because they are disinclined to believe anything. Zealots may be muggish, and Mugs zealous, in their own professional activities. Neither side has a monopoly of virtue. A man may be zealous in a good cause or a bad one; a Mug may be a saint or a cynic. Zealots dominate traditional centres of political

---

VII. MUGS AND ZEALOTS

GENERAL ATTITUDES TO SCIENCE	MUGS	ZEALOTS
Aim in research is	understanding nature surer education	power over nature national competence
Aim in education is	benefit to individual	benefit to society
Aim in technology is	improving life serving individual	transforming life promoting social cohesion
Control of technology by	social tests	economic tests
React to social problems with	ten-year research programme	quick technological fix
Long-term panacea is	education	production
Scientists seen as	hasty submissive	slow subversive
Technologists seen as	producing too much hardware	the salt of the earth
Role of foresight	to see dangers	to see opportunities
Plan research for	exploration questioning goals	pursuit of goals
SOME CHARACTERISTICS		
Admired virtues	scepticism prudence iconoclasm nonconformity	conviction boldness authority loyalty
Political connotations	internationalism 'doves' nostalgic conservatism ideal communism liberalism democracy	patriotism 'hawks' economic conservatism communism in practice fascism revolution

power, because they are more forceful. The Mugs dominate the arts and sciences for an equally simple reason: parroting of existing ideas is boring and professionally unadmired. The Mug's weakness is an aversion from action, whether political fighting or major technological enterprise; he may be too 'reasonable'. On the other hand, the most potent new ideas, social and scientific, tend to originate from Mugs.

	MUGS	ZEALOTS
Common descriptions	concerned woolly-minded	decisive fanatical
Some heroes	Mahatma Gandhi Robert Kennedy Martin Luther King Albert Einstein Charlie Brown	V. I. Lenin Winston Churchill Moshe Dayan Henry Ford Superman
PARTICULAR ATTITUDES TO TECHNOLOGIES		
Military use of H-bombs is	indefensible	conceivably justifiable
Development of BCW* *is*	indefensible	prudent
Manned spaceflight is	wasting resources	tremendously exciting
Space exploration is	for co-operation	for competition
Satellite broadcasting is		
nationally for	non-profit agency	government or private enterprise
internationally for	international agency	national or private enterprise
Ocean rights assigned	under international control	as rewards for enterprise
Wild-life conservation is	desperately important	okay within limits
Top priorities in medical research are	family planning preventive medicine	cure for cancer heart replacements
The automobile is	dangerous and obsolete	a man's best friend
Monitoring devices may	enslave us	eliminate crime
Super-intelligent computers may	enslave us	make us smarter
Mood-controlling drugs may	enslave us	make us more sociable
Use of psychedelic drugs is	tolerable, or may be discouraged	salvation, or to be prohibited
Attitudes to eugenics are that	human beings are okay	human beings should be improved

*biological and chemical weapons

Technology itself has a built-in bias towards zealotry, especially when embodied in huge industries or elaborate nationalistic projects like sending men to the Moon. The concept of the technological fix may be Zealot-like, even when applied to muggish ends. On the other hand, the new style of socially oriented engineer (Chapter 10) will be less zealous for particular technologies, especially if he is one of Herbert Hollomon's engineers who wonders if engineering is worth doing.

The democracy of the second kind, based on the higher educational system, will have muggishness built in, both because of its provenance and because its principal subject matter, the futures debate, assumes that alternative futures can be contemplated and none is self-evidently right. The muggish haven is in scientific research itself. To their great disability, the necessary specialization of the scientists induces local zealotry for their subjects, but that is true of all divisions of labour. As a corporate activity, research is as muggish as you can wish, and the quest for objectivity is a form of altruism. The American psychologist James Frank (1967) wrote:

> 'The ideology of science itself is probably the most potent antidote to the tribal belief systems that still threaten to bring the world down in flames. As this ideology—which stresses the search rather than dogmatic certainty, respects rather than condemns differences of opinions, and unites men rather than divides them—produces increasing benefits to mankind, it is bound to gain increasing sway over the minds of the people of all nations.'

The Mugs will gain power because, as Winston Churchill said of science, the new empires are empires of the mind. That is perhaps the chief reason why those who know all the reasons for being pessimistic about our technological future can yet remain optimistic. It is scarcely credible that the well-informed, research-trained, interconnected communities of the foreseeable future should tolerate men in power who want power for its own sake, who have simple-minded theories of society, or who seek to profit from the differences between men. The wind of learning and self-knowledge will topple them. Their fall will not end the search for new ways of life; it will merely allow it to begin in earnest.

Some working hypotheses (not creeds!) suggest themselves to make Technopolis habitable:

that diversity is better than uniformity; that as a social and cultural experiment one nation need be no more of a threat to others than one laboratory is to another; that there is nothing that men should not know, but some things they should not do with their knowledge; that all social hypotheses—including even grand philosophies—are tolerated as a genetic pool of ideas but none is given exclusive play.

Constructive, imaginative debate will displace those radical and conservative principles which both deny options:

'We must do it, because it's the latest thing' and

'We mustn't do it, because we never did it before.'

With such an attitude, we need not fear our ability to handle the strange powers of nuclear fusion, genetic control, or total information, but rather welcome them, as opening new territory to the robust spirit of man.

AUTHOR'S NOTE

THIS BOOK deals mainly with events and expectations at the end of the 1960s, but my anterior debt is to my father and to the public-spirited British scientists of his generation who were writing about the social connections of science in the 1930s, and who still give leadership and hope to the inhabitants of Technopolis.

I have visited eighteen countries in five continents to talk with countless research workers, engineers, politicians and governmental and international officials; also with those rare spirits who devote their careers to the sociology and politics of science. Their information and opinions about current advances provide the foundation of the book. If I were fairer to them than to the reader, each page would be peppered with their names.

I have merely documented the arguments to a limited extent, without attempting a work of critical scholarship. The growing body of academic literature on science policymaking is imperfectly reflected; in seeking to break new ground I have cut short the usual histories and analyses of institutional arrangements and have relied heavily, in that connection, upon the work of the Organization for Economic Co-operation and Development, which synthesizes the best thinking on 'policy for science'. The bibliography consists mainly of cited sources, but a small, necessarily arbitrary selection of other relevant books and reports is included.

Thanks are due to Liz Calder for research, to Lily Douglas for secretarial work, to Rena Feld for textual comments and to Donald Plunkett and staff of the Crawley branch of the West Sussex County Library, who tracked down many books.

Crawley, Sussex; March 1969. N.C.

Bibliography

[Textual references to the Bibliography are to the British editions, but for the convenience of American readers, American editions, where available, have been added in parentheses.]

ALBU, AUSTEN, Speech at Science and Parliament Conference in Vienna. May 1964.

AMERICAN ACADEMY OF ARTS AND SCIENCES, "Toward the Year 2000." *Daedalus,* Summer 1967.

AMERICAN ASSOCIATION FOR THE ADVANCEMENT OF SCIENCE, *The Integrity of Science.* Washington, 1964.

ARMYTAGE, W. H. G., *The Rise of the Technocrats.* London, Routledge & Kegan Paul, 1965. (University of Toronto Press, 1965.)

ARON, RAYMOND, *The Industrial Society.* London, Weidenfeld & Nicolson, 1967. (New York, Praeger, 1967.)

ARROW, K. J., *Social Choice and Individual Values.* New York, Wiley, 1963.

ARTSIMOVICH, LEV, Pugwash Conference, Ronneby, Sweden. 1967.

ASHBY, SIR ERIC, *Technology and the Academics.* London, Macmillan, 1959. (New York, St. Martin's Press, 1958.)

ATTLEE, LORD, Recorded interview BBC TV. October 8, 1967.

AUERBACH, ISAAC, Paper at the International Center for the Communication Arts and Sciences, New York. November 19, 1967.

BAGEHOT, WALTER, 1872. *The English Constitution.* Republished. London, Collins, 1963. (Ithaca, N. Y., Cornell University Press, 1966.)

BARBER, BERNARD, and HIRSCH, W., *The Sociology of Science.* New York, Free Press, 1967.

BARBER, RICHARD J., *The Politics of Research.* Washington, Public Affairs Press, 1966.

BARZUN, JACQUES, *Science, the Glorious Entertainment.* London, Secker & Warburg, 1964. (New York, Harper & Row, 1964.)

BEN-DAVID, JOSEPH, *Fundamental Research and the Universities.* Paris, OECD, 1968.

BERNAL, J. DESMOND, *The Social Function of Science.* London, Routledge & Kegan Paul, 1939. (Cambridge, Mass., M. I. T. Press, 1967.)

———, In *New Scientist,* vol. 28, p. 215, 1965.

BEVERIDGE, W. I. B., *The Art of Scientific Investigation.* London, William Heinemann, 1950. (rev. ed., New York, W. W. Norton, 1957.)

BLACKETT, P. M. S., Address at Royal Society. November 21, 1967.

BLAGONRAVOV, ANATOLY, In *Bulletin of the Atomic Scientists,* October 1967.

BODE, HENDRIK, In BROOKS (q.v.).

BOORSTIN, DANIEL J., *The Genius of American Politics.* Chicago, University of Chicago Press, 1953.

BORGESE, ELISABETH MANN, *The Ascent of Woman.* London, MacGibbon & Kee, 1963. (New York, Braziller, 1962.)

BORMAN, FRANK, Press conference, U. S. Embassy, London. February 3, 1969.

BORN, MAX, *Physics and Politics.* Edinburgh, Oliver and Boyd, 1962. (New York, Basic Books, 1962.)

BOWDEN, LORD, In *New Scientist,* vol. 27, p. 848 and vol. 28, p. 48, 1965.

BRIGGS, ASA, In CALDER: *The World in 1984* (q.v.).

BROOKS, HARVEY, chairman, *Applied Science and Technological Progress.* Washington, NAS, 1967.

BROWN, HARRISON, Pugwash Conference, Ronneby, Sweden. 1967.

BRYANT, LYNWOOD, In *Scientific American,* March 1967.

BURCKHARDT, JACOB, Quoted in *The* (London) *Times,* October 17, 1967.

BURKE, JOHN G., *The New Technology and Human Values.* Belmont, Calif., Wadsworth, 1966.

BUSH, VANNEVAR, *Modern Arms and Free Men.* London, William Heinemann, 1950. (Cambridge, Mass., M. I. T. Press.)

BUTT, RONALD, *The Power of Parliament.* London, Constable, 1967.

CALDER, NIGEL, *The Environment Game.* London, Secker & Warburg, 1967 (b). (*Eden Was No Garden.* New York, Holt, Rinehart & Winston, Inc., 1967.)

———, ed., *The World in 1984.* Harmondsworth, Penguin Books, 1965. (Baltimore, Penguin Books, 1965.)

———, ed., *Unless Peace Comes.* London, Allen Lane, 1968 (a). (New York, Viking Press, 1968.)

———, In *New Scientist,* (anon.), vol. 15, p. 489, 1962.

———, In *New Scientist,* vol. 22, pp. 533 ff., 1964.

———, In *New Statesman,* March 10, 1967 (a).

———, In *New Statesman,* April 21, 1967 (c).

———, In *New Statesman,* August 30, 1968 (b).

CALDER, RITCHIE, *The Inheritors.* London, William Heinemann, 1961. (*After the Seventh Day.* New York, Simon and Schuster, 1961.)

———, *Living with the Atom.* Chicago, University of Chicago Press, 1962.

———, "World in a Box." BBC TV. January 18, 1968.

CAREY, WILLIAM, In *Reviews of National Science Policy—United States.* Paris, OECD, 1968.

CARTER, ANNE, In *Scientific American,* April 1966.

CHARYK, JOSEPH, Quoted in *Bulletin of the Atomic Scientists,* April 1968.

COCKCROFT, SIR JOHN, In *New Scientist,* vol. 28, p. 366, 1965.

COMMONER, BARRY, *Science and Survival.* London, Gollancz, 1966. (New York, Viking Press, 1966.)

CONANT, JAMES, *Modern Science and Modern Man.* New York, Columbia University Press, 1952.

COTTRELL, A. H., Lectures to National Academy of Engineering in Washington. Summarized in *New Scientist,* vol. 38, p. 296, 1968.

COUNCIL OF EUROPE, Strasbourg,

Fresh Water Pollution Control. 1966.

———, *European Water Charter.* 1968.

CROSSMAN, RICHARD, Speech in the House of Commons. *Hansard,* December 1966.

CROWTHER, J. G., *Francis Bacon.* London, Cresset Press, 1960. (Chester Springs, Pa., Dufour Editions, 1963.)

CSSR (Czechoslovakia) Communist Party, Action Statement. April 1968.

CZERNETZ, KARL, 1964 Speech at Science and Parliament Conference, Vienna. In *Science and Parliament,* Paris, OECD, 1965.

DADDARIO, EMILIO, chairman, Second Progress Report on Science, Research and Development. 89th Congress, Second Session 69–209. Washington, GPO, 1966.

———, In *Physics Today,* vol. 20, no. 10, p. 83, 1967.

DAVIS, KINGSLEY, In *Science,* vol. 158, pp. 1307 ff., 1967.

DAVIS, NUEL PHARR, *Lawrence and Oppenheimer.* New York, Simon and Schuster, 1968.

DEDIJER, STEVAN, In *New Scientist,* vol. 21, pp. 461 ff., 1964.

DE GAULLE, CHARLES, French TV. June 7, 1968 (a).

———, Quoted in *The* (London) *Times,* August 26, 1968 (b).

DE JOUVENEL, BERTRAND, *L'Art de la Conjecture.* Paris, Du Rocher, 1964.

DJERASSI, CARL, Pugwash Conference, Ronneby, Sweden. 1967. Also in *Bulletin of the Atomic Scientists,* Januarv 1968.

DUBOS, RENÉ, *Man Adapting.* New Haven, Yale University Press, 1965.

DUNNING, JOHN, In *The* (London) *Times,* January 4, 1968.

DUPREE, A. HUNTER, *Science in the Federal Government.* Cambridge, Mass., Belknap Press of the Harvard University Press, 1957.

EINSTEIN, ALBERT, Letter to Italian scientists. 1950.

EISENHOWER, DWIGHT, Farewell address, Washington. January 17, 1961.

EKLUND, SIGVARD, IAEA press release. April 27, 1968.

ELLUL, JACQUES, *The Technological Society.* New York, Alfred A. Knopf, 1964.

ENCKE, STEPHEN, *Defense Management.* New York, Prentice-Hall, 1967.

EYSENCK, HANS J., *The Psychology of Politics.* London, Routledge & Kegan Paul, 1954. (New York, Humanities Press, 1963.)

FARRINGTON, BENJAMIN, *Science and Politics in the Ancient World.* London, Allen & Unwin, 1939. (2d ed., New York, Barnes & Noble, 1966.)

FASCELL, DANTE B., chairman, *The United Nations and the Issue of Deep Ocean Resources.* 90th Congress, First Session, 84–771. Washington, GPO, 1967.

FERRY, WILBUR, Quoted in *Technology Review,* May 1967.

FORRESTER, JAY, Speech at NAE fall meeting, Washington, D. C., 1967.

FRANCK REPORT, THE, 1945. Reprinted in SMITH: *A Peril and a Hope* (q.v.).

FRANK, J. D., Testimony at Senate Foreign Relations Committee. 1967.

FREEMAN, CHRISTOPHER, In *Science,* vol. 158, pp. 463 ff., 1967.

FREEMAN, C., and YOUNG, A., *The Research and Development Ef-*

fort in Western Europe, North America and the Soviet Union. Paris, OECD, 1965.

FULLER, R. BUCKMINSTER, In *Architectural Design,* vol. 37, pp. 61 ff., 1967.

GABOR, DENNIS, Remarks at International Futures Research Inaugural Congress, Oslo. September 1967.

GALBRAITH, JOHN KENNETH, *The New Industrial State.* London, Hamish Hamilton, 1967. (Boston, Houghton Mifflin, 1967.)

GALLEY, ROBERT, chairman, *Gaps in Technology—Electronic Components.* Paris, OECD, 1969.

GERHOLM, T. R., *Physics and Man.* Totowa, N. J., Bedminster Press, 1967.

GIANNOTTI, GIANNI, Paper for Fondazione Giovanni Agnelli, Rome Conference. April 6, 1968.

GILL, STANLEY, Press conference, British Computer Society. June 15, 1967.

GILPIN, ROBERT, *American Scientists and Nuclear Weapons Policy.* Princeton, N.J., Princeton University Press, 1962.

GISCARD, D'ESTAING, OLIVIER, chairman, *Preliminary Report of the Study Committee on the Creation of a European Institute of Science and Technology.* 1968.

GLACKEN, CLARENCE J., *Traces on the Rhodian Shore.* Berkeley, Calif., University of California Press, 1967.

GLASS, BENTLEY, *Science and Ethical Values.* London, Oxford University Press, 1966. (Chapel Hill, N.C., University of North Carolina Press, 1969.)

GLUSHKOV, VICTOR, In *Soviet Weekly,* November 19, 1966.

GORTON, JOHN, Quoted in *Science,* vol. 160, p. 73, 1968.

GOWING, MARGARET, *Britain and Atomic Energy.* London, Macmillan, 1964. (New York, St. Martin's Press, 1964.)

GREEN, PHILIP, *Deadly Logic.* Columbus, Ohio State University Press, 1966.

GREENBERG, DANIEL S., *The Politics of Pure Science.* New York, New American Library, 1968.

HAHN, OTTO, *A Scientific Autobiography.* London, MacGibbon & Kee, 1967. (New York, Scribner's, 1966.)

HAILSHAM, LORD (QUINTIN HOGG), *Science and Politics.* London, Faber & Faber, 1963.

HARRIS, FRED, chairman, *Report on Sub-Committee on Government Research.* 1966.

HELLER, JOSEPH, *Catch-22.* London, Jonathan Cape, 1962. (New York, Simon and Schuster, 1961.)

HELMER, OLAF, *Social Technology.* New York, Basic Books, 1966.

HETMAN, FRANÇOIS, *L'Europe de l'Abondance.* Paris, Fayard, 1967.

HEWLETT, RICHARD G., and ANDERSON, O. E., *The New World 1939–1946.* University Park, Pa., Pennsylvania State University Press, 1962.

HINSHELWOOD, SIR CYRIL, Address at Royal Society. November 30, 1960.

HOLLOMON, HERBERT, Report by J. Campbell, *Evening Standard,* March 19, 1967 (a).

———, In *Technology Review,* May 1967 (b).

HOOKE, ROBERT, *Posthumous Works.* 1705. Quoted in CROWTHER (q.v.).

HORNIG, DONALD, In *Reviews of*

National Science Policy—United States. Paris, OECD, 1968.

HOYLE, FRED, In *Physics Today*, vol. 21, no. 4, pp. 148ff., 1968.

HUMPHREY, HUBERT, 1966 Broadcast quoted in WEAVER (q.v.).

———, OECD press release. April 7, 1967.

HUXLEY, ALDOUS, *Brave New World*. London, Chatto & Windus, 1932. (New York, Harper & Row, 1932.)

———, *Island*. Harmondsworth, Penguin Books, 1964. (New York, Harper & Row, 1962.)

IRVING, DAVID, *The Virus House*. London, William Kimber, 1967. (*The German Atomic Bomb*. New York, Simon and Schuster, 1968.)

JANTSCH, ERICH, *Technological Forecasting in Perspective*. Paris, OECD, 1967.

JEWKES, J., SAWYERS, D., and STILLERMAN, R., *The Sources of Invention*. London, Macmillan, 1958. (New York, St. Martin's Press, 1958.)

JOHANNESSON, OLOF, *The Great Computer*. London, Gollancz, 1968.

JOHNSON, LYNDON B., Quoted in *Science*, vol. 159, p. 858, 1968.

JONES, F. E., chairman, *The Brain Drain*, CMND 3417. London, HMSO, 1967.

KAHALAS, SHELDON, In *Physics Today*, vol. 21, no. 3, p. 11, 1968.

KAHN, HERMAN, *On Escalation*. Hudson Institute, 1965.

———, *On Thermonuclear War*. Princeton, N.J., Princeton University Press, 1960.

KAPITZA, PETER, 1922. Letter in A. PARRY, *Peter Kapitsa on Life and Science*. New York, Macmillan, 1968.

———, Quoted in *New Scientist*, vol. 29, p. 235, 1966 (a).

———, Lecture at Trinity College, Cambridge. May 16, 1966 (b).

KAPLAN, NORMAN, ed., *Science and Society*. Chicago, Rand McNally, 1965.

KEFAUVER, ESTES, Quoted in *New Scientist*, vol. 15, p. 199, 1962.

KENNEDY, JOHN, 1963 Address at centennial meeting, National Academy of Sciences. Published in *The Scientific Endeavour*. New York, Rockefeller Institute, 1964.

KISTIAKOWSKY, GEORGE B., chairman, *Basic Research and National Goals*. Washington, NAS, 1965 (a).

———, BBC Third Programme. November 16, 1965 (b).

KLEIN, OTA, Contributions to International Futures Research Inaugural Congress, Oslo. September 1967.

KOESTLER, ARTHUR, *The Ghost in the Machine*. London, Macmillan, 1967. (New York, Macmillan, 1968.)

KOWARSKI, LEW, *An Account of the Origin and Beginnings of CERN*. Geneva, CERN, 1961.

LANG, DANIEL, *An Enquiry into Enoughness*. London, Secker & Warburg, 1960. (New York, McGraw-Hill, 1965.)

LANGER, E., In *Science*, vol. 157, p. 1533, 1967.

LAPP, RALPH, E., *The New Priesthood*. New York, Harper & Row, 1965.

LEAR, JOHN, In *New Scientist*, vol. 3, p. 14, 1957.

———, In *New Scientist*, vol. 15, p. 199, 1962.

LEKACHMAN, R., *The Age of Keynes*. London, Allen Lane, 1967. (New York, Random House, 1966.)

LEWIN, LEONARD, *Report from Iron Mountain.* London, MacDonald, 1967. (New York, Dial Press, 1967.)

LONG, EDWARD, Quoted in *Washington Newsletter,* May 1968.

LOVELL, SIR BERNARD, *The Story of Jodrell Bank.* London, Oxford University Press, 1968. (New York, Harper & Row, 1968.)

LUNDBERG, BO, In *New Scientist,* vol. 9, p. 460, 1961.

LURIA, SALVADOR, In SONNEBORN (q.v.).

MACDONALD, GORDON, In CALDER: *Unless Peace Comes* (q.v.).

MADDOX, JOHN, In *Nature,* vol. 218, p. 630, 1968.

MAGAT, MICHEL, Pugwash Conference, Ronneby, Sweden, 1967.

MALRAUX, ANDRÉ, In *The* (London) *Times,* November 18, 1967.

MARCUSE, HERBERT, *One Dimensional Man.* London, Routledge & Kegan Paul, 1962. (Boston, Beacon Press, 1964.)

MASSEY, SIR HARRIE, chairman, First Report on Science Policy. CMND 3007. London, HMSO, 1966.

———, Second Report on Science Policy. CMND 3420. London, HMSO, 1967 (a).

———, Speech at University College, London. May 3, 1967 (b).

MEDAWAR, P. B., *The Art of the Soluble.* London, Methuen, 1967. (New York, Barnes & Noble, 1967.)

MENKE-GLÜCKERT, PETER, Address to the British Association for World Government, London. April 12, 1967.

MERTON, ROBERT K., In BARBER and HIRSCH (q.v.).

MONOD, JACQUES, In *Science,* vol. 150, p. 1015, 1963.

MORRIS, EDWARD A., In SEWELL (q.v.).

MYRDAL, ALVA, Pugwash Conference, Ronneby, Sweden. 1967.

MYRDAL, GUNNAR, *Beyond the Welfare State.* London, Duckworth, 1958. (New Haven, Yale University Press, 1960.)

Nature, Article on Civil Nuclear Power, vol. 216, p. 735, 1967.

OECD (Organization for Economic Co-operation and Development), Paris. Bellagio Declaration. OECD press release. November 25, 1968.

———, *Fundamental Research and the Policies of Governments.* 1966.

———, *Japan.* 1967.

———, *Overall Level and Structure of R and D Efforts.* 1967.

———, *Problems of Science Policy.* 1968.

———, *Reviews of National Science Policy—France.* 1966.

———, *Science and Parliament,* 1964.

———,*USSR.* 1969.

———, *United Kingdom and Germany.* 1967.

———, *USA.* 1968. (See also Ben-David, Freeman, Galley, Jantsch.)

OELE, A. P., *European Parliament Document 97.* September 23, 1966.

OLDHAM, C. H. G., Lecture to Royal Society of Arts, London. March 25, 1968.

ORR, JOHN B., *As I Recall.* London, MacGibbon & Kee, 1966. (Garden City, N.Y., Doubleday, 1966.)

OSGOOD, C., and UMPLEBY, S., Paper at International Futures Research Inaugural Congress, Oslo. September 1967.

PACKARD, VANCE, In *The New York Times Magazine,* January 8, 1967.

PALOCZI-HORVATH, GEORGE, *The*

Facts Rebel. London, Secker & Warburg, 1964. (New York, Hillary House, 1965.

PARKINSON, C. NORTHCOTE, In *New Scientist,* vol. 13, pp. 193 ff., 1962.

POPE PAUL VI, *Evangelical Letter.* March 26, 1967.

PECCEI, AURELIO, Lecture to Siberian branch of Soviet Academy of Science, Akademgorodok. 1967.

PLANK, JOHN, Senate hearings on International Social Science. 89th Congress, Second Session, 77–630. Washington, GPO, 1966.

POINCARÉ, HENRI, *Dernières Pensées.* 1913. Republished as *Mathematics and Science: Last Essays.* New York, Dover.

POLLARD, ERNEST, In *Science,* vol. 157, p. 755, 1967.

PRICE, DON K., *Government and Science.* London, Oxford University Press, 1962. (New York, New York University Press, 1954.)

———, *The Scientific Estate.* London, Oxford University Press, 1965. (Cambridge, Mass., Harvard University Press, 1965.)

———, Address at Science of Science Meeting, Edinburgh. 1966.

RAMO, SIMON, New York Academy of Sciences, 150th Anniversary Meeting. December 1967.

REED, SHELDON, Paper at Nobel Conference, Minnesota, January 1965. Published in ROSLANSKY (q.v.).

REVELLE, ROGER, In *New Scientist,* vol. 17, p. 340, 1963.

RICHTA, RADOVAN, ed., *Civilisation at the Crossroads.* Institute of Philosophy, Czech Academy of Sciences, 1967.

———, ed., *Sociologicky Casopic No. 2* (Special Issue). 1966.

RICKOVER, HYMAN G., Granada lecture, London. October 27, 1965.

ROBERTS, WALTER ORR, Address at American Association for the Advancement of Science meeting, Washington. 1966.

ROBINSON, RICHARD, In *Technology Review,* March 1967.

ROSLANSKY, JOHN D., ed., *Genetics and the Future of Man.* North-Holland, 1966. (New York, Appleton-Century-Crofts, 1967).

ROTBLAT, JOSEPH, *Pugwash: A History of the Conferences on Science and World Affairs.* Czech Academy of Sciences, 1967.

RUSSELL, BERTRAND, *Nightmares of Eminent Persons.* London, Bodley Head, 1954. (New York, Simon and Schuster, 1955.)

———, *Wisdom of the West.* London, MacDonald, 1959. (Garden City, N.Y., Doubleday, 1959.)

SAKHAROV, ANDREI, Underground MS "Thoughts on Progress, Peaceful Coexistence and Intellectual Freedom." In *The New York Times,* July 22, 1968.

SALAM, ABDUS, In *New Scientist,* vol. 17, p. 515, 1963.

SALOMON, JEAN-JACQUES, "La Science et Les Sociéties Industrielles." *Les Etudes Philosophiques,* no. 2, p. 205 ff., 1966.

SCHLESINGER, JAMES, Memorandum on Uses and Abuses of Analysis. 90th Congress, Second Session, 92–944. Washington, GPO, 1968.

SCHRAMM, WILBUR, *Communications Satellites for Education, Science and Culture.* Paris, UNESCO, 1968.

SCHWARTZ, CHARLES, In *Physics*

Today, vol. 21, no. 3, p. 9, 1968.

SERVAN-SCHREIBER, JEAN-JACQUES, *Le Défi Américain*. Paris, Denöel, 1967. (*The American Challenge*. New York, Atheneum, 1968.)

SEWELL, W. R. DERRICK, ed., *Human Dimensions of Weather Modification*. Chicago, University of Chicago Press, 1966.

SHILS, EDWARD, *The Torment of Secrecy*. London, William Heinemann, 1956. (New York, Macmillan, 1956.)

SHONFIELD, ANDREW, *Modern Capitalism*. London, Oxford University Press, 1965. (New York, Oxford University Press, 1965.)

SHORT, EDWARD, Press conference on the Open University, London. January 27, 1969.

SIK, OTA, *Plan and Market Under Socialism*. Czech Academy of Sciences, 1967.

SIZER, THEODORE, In *Nature*, vol. 217, p. 698, 1968.

SKOLNIKOFF, EUGENE B., *Science, Technology and American Foreign Policy*. Cambridge, Mass., M. I. T. Press, 1967.

SMITH, ALICE KIMBALL, *A Peril and a Hope*. Chicago, University of Chicago Press, 1965.

SMITH, BURKE W., In *Scientific American*, January 1967.

SNOW, C. P., *Science and Government*. London, Oxford University Press, 1961. (Cambridge, Mass., Harvard University Press, 1961.)

SODBERG, RICHARD, In BROOKS (q.v.).

SONNEBORN, T. M., ed., *The Control of Human Heredity and Evolution*. New York, Macmillan, 1965.

SÖRM, FRANTIŠEK, Foreword to RICHTA: *Civilisation at the Crossroads* (q.v.).

SPAGHT, MONROE, In *Shell Science and Technology Newsletter*. November 1966.

SPEISER, A. P., In *New Scientist*, vol. 20, p. 391 ff., 1963.

STAPLEDON, OLAF, *Last and First Men*. Harmondsworth, Penguin Books, 1937.

STORER, NORMAN W., *The Social System of Science*. New York, Holt, Rinehart & Winston, 1966.

SZILARD, LEO, 1948 story *"The Mark Gable Foundation."* Republished in *The Voice of the Dolphins*. London, Gollancz, 1961. (New York, Simon and Schuster, 1961.)

TATON, RENÉ, *Reason and Chance in Scientific Discovery*. London, Hutchinson, 1965. (New York, Science Editions, 1962.)

THRING, M. W., *Journal of the Royal Society of Arts*. April 1966.

TITMUSS, RICHARD, In *Science Journal*, June 1967.

UNESCO, *World Directory of National Science Policy-Making Bodies*, First volume, Paris, 1966.

VAN ALLEN, JAMES, In *Science*, vol. 158, p. 1405, 1967.

VOGT, EVON, In SEWELL (q.v.).

VUCINICH, ALEXANDER, *Science in Russian Culture*. Stanford, Calif., Stanford University Press, 1963.

WADDINGTON, C. H., *The Ethical Animal*. London, Allen & Unwin, 1960. (New York, Atheneum, 1961.)

WATSON, JAMES D., *The Double Helix*. London, Weidenfeld & Nicolson, 1968. (New York, Atheneum, 1968.)

———, In *Reviews of National Science Policy—United States.* Paris, OECD, 1968.

WEAVER, WARREN, *Science and Imagination.* New York, Basic Books, 1967.

WEBB, JAMES, Quoted in *The* (London) *Times,* April 25, 1967.

WEDGWOOD BENN, ANTHONY, Speech at British Computer Society, Southampton. September 25, 1967.

———, Speech at Council of Europe, Strasbourg. January 29, 1968 (a).

———, Speech at IEE, London. May 14, 1968 (b).

———, Speech at Welsh Council of Labour, Llandudno. May 25, 1968 (c).

WEINBERG, ALVIN, *Reflections on Big Science.* Cambridge, Mass., M. I. T. Press, 1967 (a).

———, In *Bulletin of the Atomic Scientists,* April 1966.

———, Paper for the International Atomic Energy Agency, Vienna. December 1967 (b).

WEIZMANN, CHAIM, *Trial and Error.* London, Hamish Hamilton, 1949. (New York, Schocken Books, 1966.)

WHEELER, HARVEY, In CALDER: *Unless Peace Comes* (q.v.), (b).

———, In *The Center Magazine,* Center for the Study of Democratic Institutions, vol. 1, no. 3, p. 49 ff., 1968 (a).

WHITE, LYNN, Paper at American Association for the Advancement of Science meeting, Washington. 1966.

WILSON, HAROLD, Speech at the Mansion House, London. November 30, 1966.

WIRTZ, WILLARD, Quoted in Letter to President Johnson from Ad Hoc Committee. March 22, 1964.

WOODWELL, GEORGE, In *Scientific American,* March 1967.

WOOLRIDGE, DEAN, chairman, *Biomedical Science and its Administration.* White House, 1965.

ZUCKERMAN, SIR SOLLY, 1961 lecture at SHAPE. Republished in ZUCKERMAN: *Scientists and War.* London, Hamish Hamilton, 1966. (New York, Harper & Row, 1967.)

———, Lecture to the Science of Science Foundation, London. April 10, 1967.

———, Address to the Parliamentary and Scientific Committee, London. February 1968.

ZWORYKIN, VLADIMIR, In CALDER: *The World in 1984* (q.v.).

Index

Nigel Calder is well known in scientific circles and is England's foremost science writer. He formerly edited *The New Scientist* and is at present science correspondent for *The New Statesman*. He has contributed to many scientific journals throughout the world, is editor of *The World in 1984* and *Unless Peace Comes*, and is the author of *The Environment Game*.